Heinzwerner Preuß

Atoms and Ethics

Heinzwerner Preuß

ATOMS AND ETHICS

On Ethics and a Worldview Implied by the Insights of Science

LIT

This book is printed on acid-free paper.

Bibliographic information published by the Deutsche Nationalbibliothek
The Deutsche Nationalbibliothek lists this publication in the Deutsche Nationalbibliografie; detailed bibliographic data are available in the Internet at http://dnb.d-nb.de.

ISBN 978-3-643-90636-6

Original title: Das Atom und die Ethik
ISBN 978-3-643-10104-4

Translated by Publisher

A catalogue record for this book is available from the British Library

Klosbachstr. 107
CH-8032 Zürich
Tel. +41 (0) 44-251 75 05 Fax +41 (0) 44-251 75 06
E-Mail: zuerich@lit-verlag.ch http://www.lit-verlag.ch

Distribution:

In the UK: Global Book Marketing, e-mail: mo@centralbooks.com
In North America: International Specialized Book Services, e-mail: orders@isbs.com
In Germany: LIT Verlag Fresnostr. 2, D-48159 Münster
Tel. +49 (0) 2 51-620 32 22, Fax +49 (0) 2 51-922 60 99, E-mail: vertrieb@lit-verlag.de
In Austria: Medienlogistik Pichler-ÖBZ, e-mail: mlo@medien-logistik.at
e-books are available at www.litwebshop.de

Author's Introduction

This book is not the usual kind of book about matter.

As a physicist, my goal is rather to present the implications that the current state of scientific understanding (quantum theory) truly holds for all of humanity, if all areas of life are taken into account, and how certain these conclusions are.

The book is the distillation of nearly 50 years of work as a researcher and university teacher in the field of theoretical chemistry. This topic had already piqued my interest when I worked at the Max Planck Institute for Physics under Professor Heisenberg, first in Göttingen and later in Munich (1952–69), and I continued my engagement after my appointment in 1969 to the chair in theoretical chemistry at the University of Stuttgart, an appointment that was part of the founding of an institute.

Only after my retirement in 1993 was it possible for me to take up these themes more intensively, themes that I had come in the meantime to regard as crucial. In the beginning, I did not foresee the implications presented here, not could I have guessed them. They arose definitively and incontrovertibly solely from the research and its methods.

From this perspective, the results and conclusions are neither a proximate work of human hands nor a revelation, as the implications must – regardless of preconceived notions – necessarily be recognized in the cosmos and deduced from it by inquiring persons, once the methods of research had reached the necessary level.

One concern of mine is to declare again that no ideology or view of the world that is claimed – or stipulated – to be true is presented here, but rather insight into nature that was worked out ad prepared by many scientists, at times against their own expectations. I am tying this together with the hope that the book will be approached with the least amount of preconceptions and as impartially as possible. Because the implications of physics change our consciousness and our relations to other people.

Ein logischer Schluß, der zur Sinneswahrnehmung im Widerspruch steht, ist unannehmbar. Ein Grundsatz, der mit der experimentellen Sinneswahrnehmung nicht übereinstimmt, ist in Wirklichkeit kein Grundsatz, sondern ein grundsätzlicher Fehler. ... Es genügt nicht, die Beobachtungen nur auf eine bestimmte Weise anzustellen. Man muß sie vielmehr unter den verschiedenen Umständen wiederholen, damit die wahre Ursache der Erscheinung mit Sicherheit ermittelt werden kann.

Albertus Magnus (Graf Albert von Bollstädt). 1193–1280.

A logical conclusion that contradicts perception is unacceptable. An axiom that does not agree with experimental evidence is in fact no axiom, but instead a fundamental mistake. ... It is not sufficient to engage observations in one particular way. The must rather be repeated in varying circumstances so that the true cause of the phenomenon may be discovered with certainty.

Albertus Magnus (Count Albert of Bollstädt), 1193–1280

Table of Contents

Summary

To obtain insights into ourselves and the world that are generally valid, unvarying, verifiable and self-confirming – and even these distinctions should be questioned – it is first necessary, in my view, to define the mental instruments being offered for use, and to subject them to critique, particularly concerning their ability to attain the goal just described.

It is astonishing that these considerations and investigations have hardly been undertaken previously; for that reason, they will be given full consideration in this book. In that consideration, the truth values of the natural sciences and the humanities will be up for discussion, and by this process brought to clarity. It is only now possible to define our knowledge of the world (insight) without coming to contradictory results, which – as the considerations show – can no longer be contradicted, only extended (relative truth). Because the characteristics listed above enable such a proposition for the first time, they thus define the concept of truth!

Without going into details that will be presented later, it is shown that thinking in models, speculation, hypotheses as well as holding things to be true (belief) are not paths that lead to true insight. Only the integration of experiences, which are individually tied to a symbol, which is then bound into a mathematical structure (theory) can guarantee a path to truth. This approach incorporates general experience and the requirement of science that everything that takes place in this world is tied to changes in the movements (probabilities) of the smallest particles of matter, and for that reason – necessarily – everything in the cosmos must be traced back to them! Thus the key distinction between the self and the cosmos is dissolved, and becomes just a question of the complexity of the systems involved. Matter, through its structure of the smallest particles, is omnipresent in the universe.

Theory now tells us that these particles of matter must be acausal in their behavior (Heisenberg's uncertainty principle). Thinking further, that means – necessarily – that there can be no creator of the cosmos – there is thus no God! These conclusions will be discussed in greater detail later. It is a matter of conclusions (insights) that are binding for all people, insofar as they are seeking for truth.

These insights, which are a part of wave mechanics, in fact lead much further. Theory further shows that influences exist among all systems of the cosmos that are comprised of the smallest particles, which can be proved, and that these influences are almost independent of space (non-locality) and can take on an enormous variety of forms and properties, so that it is possible to construct an ethics that can be applied to these complicated systems. Further, the basis for a cosmic conscience can, according to theory, not be ruled out! All of these aspects form the contents of this book.

That means: Only a society that is freed from belief can incorporate an ethics for all persons, an ethics that is in harmony with the truth, because the cosmos – via the characteristics that have now been recognized – demonstrates this path, also through theory. This ethics concerns matter's behavior toward itself!

The role of evolution will also be demonstrated later, and specifically applied to humans, as the developers of theory. This leads to a new approach to the question of the structure of global society (for example, in respect to means of payment, energy production, or genetic technology).

Finally, the propositions of theory concerning truth (mathematical relationships, laws of nature) show that all mapping of symbols of experience in a network that is free of contradiction arise in the end from a continuity of connections, in the form of the trivial fact that

$$0 = 0 \tag{1}$$

in which one side of equation (1) yields a reformulation that regulates the mapping of the symbols and leaves the value of zero unchanged. This attribution corresponds to the relative recognition of truth, as noted above. This, too, is a theme of the book.

Non-causality means that the cosmos is a product of coincidence, and experiences are evidently so constructed that in the end they correspond to a cosmos of nothingness ("divided nothingness"). The complicated systems in the universe eventually recognize the truth, as far as this is possible, and are at the same time confronted with the necessary ethics and morals.

With this, death also take on a new meaning and represents the end of a largely isolated, very complex structure of matter, one whose fundamental elements thus remain unchanged, but whose systemic complexity rules out its repetition, particularly because the changes in motion during its existence, which shape its individuality, must be taken into account.

Because the acausal behavior of the smallest particles only allows for statements of probabilities, the behavior of matter is thus similarly shaped, and shows now "material" behavior. Only larger systems – as wholes – demonstrate pseudo-causal behavior, which the theory also predicts!

With that, the basic idea of this book is demonstrated, which is, in the end, that as far as possible and without previous consideration implications should be drawn from (scientific) relative truth. That has not yet happened, but every person has a right to experience it; but to date people have had to experience much unnecessary suffering. The truth leaves no room for fanaticism or any form of exploitation.

I justify the insistent style of presentation in this book, and the frequent repetition (if at times in other words) with the importance of the insights expounded here, whose implications go far beyond what our contemporary society considers true.

Chapter 1 – Introduction[1,2]

Since the development of scientific thinking, as it developed out of philosophy, there has existed a deep misunderstanding between the natural sciences and the humanities, or, more precisely, between physical and mathematical means of gaining knowledge on the one hand, and philosophical observation of nature and speculation on the other.

The main conclusion of this misunderstanding is that the physical approach allegedly cannot make any contribution to knowledge of the world, as physics claims to do by its nature. Society finds it completely absurd that research in the natural sciences could lead to results that raise fundamental questions concerning morality and ethics, or even call the fundamentals of religion into question! For many people, such thoughts are nearly ridiculous, and reveal the ignorance of the person posing these questions. Presumptuousness on the part of the natural scientists is often seen as well.

This misunderstanding is largely a result of history, as philosophy has a longer and deep-seated tradition in human intellectual history, and the view that real knowledge of the world can only be reached through pure thought (without mathematics and experiments) has prevailed.

The very fact that all technical progress is based on scientific results also devalues physics in this sense, that it could be seen as a means to universal insight into the cosmos. Philosophy does not have an analogous external effect, if one ignores temporary spiritual currents in society. Religions, particularly the Christian religion, have often promoted this misunderstanding.

1 The present work is partly based on the author's pervious publication, Mensch und Materie, Stuttgart: Presse- und Informationsstelle der Universität Stuttgart, 1987.

2 See also Preuß, H., Materie ist nicht materiell, Vieweg-Verlag, 1997 (Springer Verlag 2000).

Furthermore, society's religious beliefs confronted scientific thought, and stood, at times, in hostile opposition to this approach to insights into the world.

With the aforementioned misunderstanding as a catalyst for certain tension between the humanities and the natural sciences, barely recognized and impassively accepted by society, it has over time received a dramatic aspect: Technology's effects are not only underestimated – because they are, after all, based on the natural sciences – but also that the connections and effects in nature that have been discovered by physics in recent decades, and which apply to the entire universe, led to implications that go far beyond the implications that could, with all due qualifications, have been derived from philosophy!

These ramifications are principally concerned with religions and questions of morals and ethics. Even the structure of our society is thus called into question, and in a revolutionary form!

It appeared that the previous conceptions, traditions and indolences had grown into an existential threat to humanity, and that was because the meanings of the results of physics were misunderstood, underestimated, or simply not acknowledged. Further, these dangers were underestimated or not recognized for lack of understanding. The role that the existence of a means of payment plays in these questions should initially be deferred, but the results of physics occasionally have a certain connection with that as well.

All of these aspects will be taken up in what follows. But first it is necessary to clarify and define the terms with which we would like to think and must think, if the above questions are to be posed.

Chapter 2 – Origin and review of terms and concepts

The development of insights into the connections among our experiences, which we originally confront mostly uncomprehendingly, and the drawing of conclusions from the insights thus won, are two separate processes, about which one can initially only say that the second step may follow from the first. Phrased that way, it sounds trivial. This impression is, however, mistaken, because things are not so simple, and the environment plays a decisive role. As long as no further description of how the insights into the connections were obtained is offered, it is not clear whether the conclusions are subject to further obligations, the statements above cannot be evaluated as trivial. That also holds true for possible constraints to which the attainment of insights is subject. In this sense, linkages between the two steps emerge, which can do much to fog up the initial clarity. The apparent conclusions are no longer definite. Gaining insights can degenerate into arbitrariness, and so these must be codified (to say nothing of revelations). Everything is possible there! Even when the opposite is often claimed.

The current work takes reaching clarity in this regard as its task, and it is presumed that these two steps – in whatever fashion – comprehensively describe persons' behavior toward their interior and exterior world! In this regard, the structures and realms of the senses, and the effects of memory (recall) play an important, but not fundamental, role. Contact, both factual and mental, must take place with the external and internal world, so that the possibility to reflect in the most general way can exist at all.

I have deliberately had to keep the preceding explanations vague because to reach a clearer and more focused treatment of the issues at hand, our language – especially the formation of concepts – must be subjected to further clarification. In other words: The terms and concepts must be sufficiently defined so that it is even possible to speak and write about the pending problem, and a minimization of possible misunderstandings (mis-interpretations) and ambiguities should be striven for. We must agree about what we are talking about!

Several examples shall serve to demonstrate what I mean. What does a journalist or a politician imagine when using the term "model"? It is a term that plays a not insignificant role in gaining or describing insights. What, for example, does a philosopher mean when speaking of a "theory"? What is a philosopher thinking, when one says that a thing is "fairly theoretical"? Do we all agree on what "belief" really means? There are communities of belief, but no communities of theory! What is the difference between speculation and hypothesis?

Eventually, we must ask ourselves what we mean when say that we have understood something. The answer to this question brings in mathematics for clarification, and its meaning for how we gain insights (mathematics, and not merely calculating).

But why is it necessary to make such clear differentiations, and why should mathematics play such an important role in this process? Is it related to our language? All of these topics must be clarified before we can take the first step, as mentioned above.

The fuzziness of our casual language can also have advantages, particularly for communicating moods, feelings or aspects of experience to other people; at those times one is not aiming to gain insights, at least not immediately. That constitutes precisely the great value of our languages, because their imprecision enables the freedom that we need in our world of feelings! As such, language in general is the medium of communications for making connections of every kind. By contrast, our common experiences call for precision, so that the results allow for clear and unambiguous communication from one person to the next!

It must be noted here that quoting other thoughts and works, however necessary in academic endeavors, can also lead to ambiguities, which then are then reinforced in general opinion, sometimes precisely because of the lack of clarity in the terms used.

It is not uncommon to hear or read scientific, but also particularly humanistic, explanations that do everything other than convey insights, that are instead collections of knowledge that offer no insight, especially when the terms in use are not generally defined and accepted. It is a precondition that everything from this side is universal, and usually clear! This is usually a matter of habit and tradition, that is, a certain intellectual convenience, which also in part, even if not immediately obviously, contributes to the widespread repudiation of mathematics.

This corner of our consciousness is also the source of religions, and in particular their later developments in which they deprecate the sciences, particularly the natural sciences. In this sense, religions also have a distorted relationship to matter, which in turn results from the aversion to deal with clearly defined terms.

Before we come to the first step, however, we should now clarify and define a number of terms that are necessary to attain insights and knowledge about our experiences with our world. Here are certain historical aspects, insofar as they are known.

We may assume that very early humans considered the natural world around them phenomenologically. After an increasingly strong division between emerging self-consciousness and the rest of "not-me" became apparent in the course of development, "seeing" became the introduction to an analytical process. We should understand the term "seeing," as including contact from all of our sense organs, with the help of the perceptions they enable, in regard to the internal and external conditions in all of nature. The structure of the sensory system is carried over into the structure and classification of the internal and external conditions; that is, into all of nature. A differentiation of impressions is taken up, the initial experience occurs as the sum of experiential areas that can also overlap and overlay one another.

I have described the situation – as I see it – so broadly and extensively because I attribute a great and decisive meaning to the beginnings of human inspection of the environment, because at that time a path was already taken whose influence on human development can still be clearly seen today and which is still having effects.

I want to point out this situation in particular because the alienation that arose between early humans and the "world," as they recognized the effects of non-human powers in nature, certainly led to a feeling of powerlessness, and thus to a generalized fear. As a result of this development, early humans attempted to interpret these powers as the expressions of beings that were similar to humans, but bigger, stronger, although also often burdened with their own weaknesses. Out of this, idealizations soon emerged, which are typical for the weaker person who asks for goodwill from these beings.

In this fashion arose the gods of nature, in whom people believed; that is, in whom people showed trust (but also feared) and whose existence people considered real! Belief must thus be defined as "consider real on the basis of internal and external experiences (insofar as they are possible)". In a certain sense, this is a matter of a belief in an authority, as we recognize it today – slightly changed – in star worship, which, in contrast to knowledge, derives its existence only indirectly, from personal motives or from unsubstantiated inspirations (revelations).

From this arose the "nature religions," with their gods for the various realms of experience. In this portion of our history, no clear distinction between the world of the gods and the world of humans was possible! Thus at the dawn of humanity there were already various nature religions in the many different groups of humans, all with differing structures of belief. Belief divides and separates people, and continues to do so today! That does not seem to disturb the majority, particularly because everything has developed slowly, and because it has yielded (provisional) advantages. Nature religions thus became organized religions – state religions. Along the way came divisions, which one can encompass with the term "sects."

People differ in their various forms of considering things real, and this contains – according to human mentality and especially collective mentality – the tendency to haughtiness toward other beliefs, rejection that goes as far as hatred and fighting against people who believe differently, especially when certain members of one's own "community of belief" call for it. The distance to absurd claims of being chosen is then not very far at all, with all of its criminal consequences. Most of humanity's wars were – in the end – wars of belief, terrible and vengeful! If overreaching possessiveness is added to the mix, the common thread of human history of the last millennia has largely been identified.

It has often been pointed out that behavior toward the gods certain similarities with behavior toward parents, or at least can be strongly shaped by this behavior. In this regard it can be suggested that adolescent aspects play a role, which would of course explain a great deal such as the need for absolutes, the demarcation from other religions, and the overemphasis of one's own position without much inquiry. In the beginning, humanity is also an adolescent, and, as we know today, many people only find their way out of adolescence very late if at all. It is unfortunately necessary to recognize that a large share of the media support this phenomenon.

Even the search for authorities that accord with one's own notions is a part of this psychic spectrum. A global example of this process is the invention and construction of stars in the media, politics and sports.

The interpretation (not understanding) of natural processes (and societal events) via gods answered this need. Subordination and veneration at the cost of self-development were tied to the process.

Much more can be said on this topic; just a few of the most important aspects that have been widely researched are set out here. The role of the intermediaries, the priesthood, has been left out, although their appearance and dominance cannot be overlooked when a situation like the one described is at hand. I am much more concerned with the connection between the way humans behave toward nature in general, and the implications that follow almost automatically, or the conclusions that people draw.

The implications that arise from the development itself are tragic. Monotheistic religions inevitably developed, which in recent centuries led to terrible and often beastly struggles, whose end is still not in sight. As could only be expected, "consolidation" of belief gave rise to "holy books," although very few know or surmise who wrote them and which base motives led to changes in the texts before the "priests" of each group presented them to the believers as inviolable wisdom and declarations. That made them into God's word! And thus was God identified as the creator of the world!

On this basis stand, with few exceptions, the humanistic studies of the West, which are counted among the greatest intellectual achievements of humanity. They rightly characterize and define Western culture. Philosophy's engagement with the world, however, has led to very different conceptions. It is true that religion is generally assumed, but "considered to be real" as a belief is not the original way of thinking in the humanities, which are much closer to the concept of intellectual freedom. Otherwise, philosophy would never have arisen!

The ambiguity of a personal "structure of thinking" in philosophy lies, as we will show later in more detail, in thinking in the realm of models, hypothesis and speculation and attempting to find solutions in this realm. Additionally, these terms are used without definition, so that endless discussions became possible, discussions that also integrate psychological moments. This should by no means be seen as denigration of philosophy, because it developed at a time when it was not possible to think in terms of the scientific method, even if occasional foreshadowing and hints of it can be read.

Philosophy's strengths and services comprise attempting to use the possibilities of thought in a given time to their utmost to see the intellectual existence of humanity and its works in a certain context, and on this basis to examine its meaning all the way through to questions of morals and ethics. In regard to this context, no agreement could be reached, which drastically limited philosophy's effects on society. Nevertheless, it is still worth recommending that young people, especially if they wish to devote themselves later to the natural sciences, begin with the study of philosophy and mathematics.

Chapter 3 – Organization of the natural sciences

Belief stands in opposition to knowledge; this becomes particularly clear when knowledge transcends simply forming opinions and considering something correct. How, precisely, is knowledge differentiated from belief, which we have already carefully described and, in the end, must view as an intellectual short-circuit? Why was it possible, in the course of development, for a different form of engagement with the world to emerge, even though belief is generally so comfortable?

In point of fact, a large portion of humanity persists in this state of "considering to be real," at least in questions of their understanding of the world, and the hold to this notion with a fanaticism that is typical for the state. Among these people it is also possible to find so-called educated persons, from whom one would expect more mature behavior, not least because today there so many sources of information are available for thinking about the world and which approach the problem from numerous points of view!

This "considering to be real" has found a particularly perverse variant in politics. In the political arena, attempts are made to portray it as a historical truth in order to justify decisions and to implement the implications. These assertions, which are considered to be real, should then be recorded in the history books. The helplessness of historians, almost without exception, is thought-provoking. Here belief is in close contact with lies, which, furthermore, is also observable in finance and advertising.

In developmental history further shows that simple, authority-centered "considering to be real" – which surely features many childish aspccts – arouses pleasant feelings because it avoids significant intellectual exertion, particularly when one is in agreement with a large number of sympathizers.

Only an almost vanishingly small number of people commit themselves to knowledge, and they thus have little influence over the majority, who often react aggressively to this commitment. This is as true for politics as it is for the media or, for example, sports, the existence of whose stars constrains thinking, or can focus it in certain directions!

Let us assert again: The knowledge that has been handed down in its entirety we call science. It constitutes a methodical handling and ordering of reality that aims for insight. That means that the requirement for verifiability of results (validity requirement) is tied to the method. Here we clearly recognize the connection with the first point touched on above (developing insights). This relationship to the world, to the cosmos, became necessary because it was increasingly clear – in the course of the historical development of consciousness – that the entire cosmos is comprised of one thing, which people called "matter," and which appeared in various states, although it was shortly recognized that this was always one and the same "substance"!

With that, we have defined the natural sciences as value-free science, in contrast to the humanities, which are also sometimes known as the judgmental sciences.

That further makes clear that only a value-free science can contribute to the topic of the first point! The second point (drawing conclusions) appears to have more to do with the judgmental sciences, but that cannot actually be determined as long as the question of which premises such sciences derive from remains open. If, by contrast, the results of the natural sciences are value-free; that is, they exist independently of humans but are valid for all people and apply to all of them, then judgments are no longer taking place, and thus the conclusions are also universally valid and applicable. They are invariant vis-à-vis persons. They are recognized by people, but people cannot influence them. Instead, they can only adjust themselves as far as they are able and behave accordingly.

We will examine this point more closely later, particularly because it occupies a central position in these reflections.

The premises of the humanities are manifold. They include all of the possibilities of our intellect, and, correspondingly, these premises are often contradictory. As a result, there can be no unique, substantiated conclusion about goals. We learn ever more about ourselves, but reach no unambiguous insights about the world as a cosmos, which is in any event often not the goal of the inquiry. How, more precisely, does the other way appear?

To answer that, we must first determine that everything, absolutely everything, that happens in the world is tied to changes in "matter." That means, if we want to gain insight into the connections between our experiences, we must engage with the "matter" at hand, this substance that is the only thing that appears in the entire cosmos and confronts us in our experiences in its various forms. We ourselves are also built upon this bedrock, and the question arises of how far engagement with "matter" will lead. The question can also be phrased as: Is the realm of experience, with which "matter" is tied, so furnished that it covers all of our possible experiences, i.e., does it include reality completely? Reality as the sum of all experiences, whereby of course the formulation of connections can expand our realm of experiences. This is because we gain experiences not only directly from our senses but also from intermediary devices!

The scientific method brought clarity for the first time to how we process experiences. That is to say, all of our experiences are, fundamentally, initially single points of information. That does not mean perceptions, but "registered" points of contact with our environment; even experiences with ourselves must of course be tied together with full consciousness and saved. These points of information – better, pieces of information from a very small realm of experience – have not yet been places in context with other experiences. In fact, we know of many historical examples in which initially many experiences were collected, without knowing (recognizing) what united them – as they perhaps (or certainly) could (must) have related.

A crucial example is light and matter. Experience showed at first that light was produced by matter, but also that matter can absorb light. What does that say more precisely about the relationship between light and matter? How should we understand these experiences?

But what exactly is understanding? Following the explanations set out above, it can be conjectured what might be meant. It is not possible to leave the experiences as a sum (collection) of points of information; that is unsatisfying for a thinking person. For that reason, one tries to find "starting-points," from which the largest possible number of individual experiences can be derived (deduced); that is, arise as a logical consequence. If, for example, two or more apparently unrelated points of experience suddenly lead back to a single "starting conjecture" (postulate), that can be a real eureka experience, because then we are looking, so to speak, behind the curtains of our original experiences that appeared to us without an inner relationship, without a connection. We may call this process "understanding," and also define it that way.

Thus the process of understanding is comprised of capturing as many experiences as possible that initially seemed unrelated with the smallest number of starting conjectures (postulates) that the experiences lead back to.

That leads immediately to the conclusion, that the better we have understood something the greater the number of experiences (areas of experience) can be accounted for by the least number of postulates that the experiences lead back to. In other words, the larger the realm of experiences overall and the smaller the number of necessary postulates, the better the experiences have been understood, and thus more soundly grounded are the insights into the structures of our experiences. It is almost unnecessary to say that these postulates are indispensable for our understanding! We call this fact "relative truth," in that by extending the realm of experiences that we have understood, the earlier experiences and understandings arise as special cases of the new, larger realm of experience (realm of understanding). Increasing truth does not contradict the elements of its development! Yes, we could go still further and conclude that a person with no postulates about the experiences that he or she intends to understand cannot reason. Postulates are not the finding (insight), but they make them possible.

In the natural sciences, we thus speak – in contrast to the dogmas of religious communities, which are a matter of belief and must be accepted by the members as such – of postulates. The acceptance of dogmas defines, so to speak, membership in the community of belief. We should, however, be clear that the postulates do not come from the outside, but arise from the experiences in that they must of necessity be introduced so that our thoughts about our experiences are in accordance with them. They are, in a certain way, a part of our experience and illustrate the structures of thought that are available in the cosmos. Thus the postulates are challenges to science, so that the measurements that they work with show a real and unambiguous relationship with experience! These descriptive measurements are also unambiguously defined and can be mathematically formulated!

We have thus described the theoretical methods of the natural sciences, particularly physics (theoretical physics) more precisely. And there is no doubt that, building on such a cross-linked system, we may expect invariable and substantiated results. Later we will often go into this set of facts, as well as addressing the aforementioned measurements more precisely.

The postulates are not provable, but they "prove" themselves, insofar as they demonstrate the common elements of the original individual experiences, and thus build understanding about our experiences – they provide insight into the world. They constitute the matrix for the structures of the thinking that we use to confront the universe! We are now in a position to encompass and describe all of nature mathematically, that is, theoretically. With this step, we will reach the maximum of our insight into nature, substantiated and verifiable; nothing must be "considered" to be real!

In the natural sciences, one thus means that understanding the connections in our complete experiences is "truer," the fewer postulates (though they can never be completely done without) are required and the greater the realm of experience that one has captured with the postulates.

In this sense, truth is relative, which, however, holds true for the entire realm of experience; I consider this to be a fundamental advantage. Furthermore, one should be aware, first, that there can be no ultimate truth, and second that a complete realm of experience whose structure one has identified – that is a realm of experience in which all processes within it can be traced back to a small number of postulate – can be seen as theoretically described, indeed, it must be seen that way. That is because nothing more can be done for understanding once it has been traced back to the smallest possible number of postulates! This interconnectedness is what we mean by "theory." In other words: Theory is the irrefutable description of all processes, and that can – in the economics of thinking – only be accomplished with the help of mathematics (which we will examine more closely later).

By "complete experience," we are understood to mean experience with "matter," which comprises the entire universe, as we know from scientific deliberations, and from which we must proceed! Above, we have already considered this topic, and here we should additionally note that in regard to "matter," complete inclusion of nature has succeeded, because no experience contradicts the universal theory of "matter." Mathematically formulated connections encompass all systems ad structures of "matter," this material about which so much more will be said!

Relative truth, in a scientific sense, does not however mean that the content of theoretical conclusions for the entire realm of experience can be changed, as that is not possible because theory is a mathematical representation of experience! We may also say that theory always applies for the realm of experience, for the realm that is understood. The relative part – if one wants to call it that at all – arises because we can deduce further, new experiences from a theory, which can then in fact take place when we attempt them. Thus a theory's realm of experience is recognized as even larger, which speaks in favor of the theory, and the overall realm of experience becomes clearer and more transparent.

In the end, we thus expect that the previous structures of our understanding remain intact, but the aforementioned integration is extended. Before people attained this insight into the process, there were a number of historical phases of understanding, which we will now analyze more closely, the more so as this approach leads to a better understanding of the role and meaning of theory (of scientific truth).

There were situations in which new experiences took place (improved possibilities for observation), experiences that could not be understood with the previously "understood" realms of experience; that is, they were not described theoretically. In this case, there really was a new situation, which could lead to a discussion of the assumptions that had previously been in use. From a historical point of view, several realms of experience were at hand.

In this case, the first reaction of (unsettled) persons is often characterized by thc formation of models and hypotheses. In the category of model-building, we also include the attempt to generally extend understanding that has already been gained and "secured" to the new, not-yet-understood area. That is the wrong path, as has been shown time and again, because a model is, at best, a transitional state! This leads to difficulties, as several "understood" areas are just formally combined, without a superstructure, which leads to contradictions.

Ambiguity remains; nor is it clear how to proceed. In the development of models notable intellectual achievements are often realized, and brilliant ideas presented!

One widely known example is the atomic model proposed by Niels Bohr in 1913, at a time, that is, when wave mechanics (theory) was not yet developed (superstructure)! The model brought together the elements of the then-emerging quantum theory with ideas from mechanics and electrodynamics.

Of course people knew at the time that the model contradicted electrodynamics in certain points, but it also became apparent that the model delivered results concerning the spectrum of the hydrogen atom that were in accord with experience and which had previously not been understood at all. At the same time, its predictions about the spatial construction of the atom contradicted experience. This situation is typical for a model!

When the theory – wave mechanics (quantum mechanics) – was later presented, all of the properties of the "matter" were easily produced, among them the properties of the hydrogen atom, and without exceptions!

Unfortunately, Bohr's model of the hydrogen is still taught in our schools, and the impression is often given that it is our latest knowledge. This leaves the population with a false and outdated view of the world.

There are numerous possibilities for moving away from thinking in models. For one, people may attempt to find out whether the new aspects are already contained in the realm of what is already known and explained, and had to date remain unnoticed. (One such case will command our attention later, as it deeply disrupts our understanding of nature.) In the case of Bohr's model, this approach failed.

Another approach is the discussion of hypotheses that already take the position that for the new realm of experience new assumptions must be discussed as an extension (superstructure), which naturally means the "transition" from the known realm must be examined. These may be called "trial theories," which one considers in the framework of the hypothesis.

Indeed, there usually is speculation – up to a hypothesis, which is then checked against experience time and again. This can at times lead to a type of model creation.

In the end, these considerations (supported by experiments) take on a consistent form regarding what an understanding (theory) of the new experiences could look like, until eventually a contradiction-free system of postulates – and their implications – has been achieved, a system that encompasses not only the new realm but also, of course, the old one. In other words: people have understood more than before; a theory has been formulated, one that describes a larger realm of experience with the least possible number of postulates.

In this process, the fact that the old theory is a special case of the new theory is often overlooked! That must always occur, because we have defined theory in this particular way; that is, the previous understanding remains intact (special case), and science experiences neither an "upheaval" nor a "revolution," as is often claimed for the benefit of the media.

If, for example, premise that correspond to the description of large particles of matter are entered into the equations of wave mechanics, we obtain the well-known propositions of classical mechanics.

Another example comes from the theory of relativity, which likewise merges into well-known mechanics, if the objects are moving at speeds that are significantly less than the speed of light (300,000 km/sec). We also obtain propositions from the general theory of relativity that correspond with the usual experience of space-time if we consider small realms of space and move within them, in order to explain something more simply.

In the course of this, it is important that a theory always permits quantitative conclusions about experiments (as these have led to its determination), so that theory and practice unambiguously bled into one another, because all of the experiences that can be had are reflected in the theory, and vice versa!

On this basis, it would be possible to rethink the postulates, in particular, whether their number could be reduced again by showing that two postulates are not independent, but rather are bound together. That means that one postulate can stand for both of them, so that the overall number of postulates is reduced, thus leading to a relative increase in the truth content of the theory.

In this context, I refer again to points made above; namely, that truth in natural science – formulated as a theory – can never be overturned, but rather only enlarged and deepened by human achievement, in which what was previously attained always serves as a solid starting point for further consideration and, possibly, as I will show later, can even lead to ethics and morals. A universally valid theory – if it includes all of the "matter" in the universe – allows for little leeway (which will also be a later topic).

In natural science, particularly in physics, a conception of the world, an insight into the world is not given *ab initio* or prescribed; instead it must be worked out by humans. The prize is the certainty of insight won, insight that is tied to experience, that is not simply a personal personal perception because it can be generally repeated and is defined as discussed above. That means that this scientific insight (insight into the connections of our experience) must be seen as a hard-and-fast chance for drawing conclusions; that has nothing in common with "considering to be true," because it is a matter of directly gained insight and knowledge about our experience! They are provable ad tied together, as demonstrated above. Ambiguities are impossible in matters of natural science. Truth in natural science has nothing to do with the term "truth" as it is used in religions. For that reason, I recommend differentiating between the two terms, "for the sake of the truth;" it would make a great contribution to clarity!

I have taken up the structure of scientific research and the meaning of its results so extensively and so often, and repeated it in other words, because far too little is known about it in our society, and this topic is practically not addressed at all in our schools. Even the fundamentals are not addressed, that which we mean by "gaining knowledge." These are in fact fundamental questions and problems of our human existence, at least in matters of thinking about the world, and its consequences are unimaginably far-reaching.

I think that it is a catastrophe over the long term that the majority of humanity believes more-or-less fanatical assertions, that other people have told them and which are supposedly of divine origin, although there is no proof and these assertions thus belong to the realm of belief. At the same time, the assertions differ among the religions, and the moral and ethical demands diverge from one another as well. The consequences in politics and for relations among communities of belief have been devastating. In the end, it is a matter of incredible credulity, which probably derives from an archaic trait of humanity!

All of this takes place before our eyes, while a vanishingly small minority makes the effort to use all of the available possibilities to illuminate, in a substantiated way, the structures of our world – that is, the "matter" – to which we also belong. Unfortunately, they have had little success.

In addition, it must be noted with concern that very many consequences of the natural sciences, together with capital, are turned against people, which has over time increasingly taken the form of general destruction. As long as a means of payment exists, the negative effects of this development will not be kept from humanity. Seemingly positive processes are concerned with the mitigation of some damages, but without long-term success, because the overall developments continue; they are driven onward.

The financing of immense projects (e.g., genetic technology, atomic energy, weapons technology) in the end speeds up this development. Certain "advantages" for minorities are the price of uninhibited "progress" toward a downfall, because these advantages deceive. One increasingly has the impression that the majority of people have become accustomed to this way of life. It would be wise to demand wide-ranging enlightenment, to perhaps yet find some ways out, but there has been no success in spreading the insights of natural science – understandably and clearly – because the public lacks the foundations, and certain circles are doing everything to ensure that circumstances remain as they are. One sees everywhere the false notion that only religions are responsible for morals and ethics, and that they proclaim the "true" conception of the world. The natural sciences, in this view, have no access to these questions because they approach things in a fashion that is very difficult to understand and too mathematically, and that they are only the concern of a few specialists. One of the greatest and most consequential errors!

For that reason, I shall address theory even more closely than before, because it characterizes natural science, from which we – as will be shown – may recognize our true position in the cosmos and the behavior that results from it. People who are still imprisoned in the old thinking will consider this conclusion "boastful and vain," but contemporary science has reached this breadth, and it is amazing that even in scientific circles these facts have not yet found entrée, so strong, it seems, are tradition and dependence.

Chapter 5 – Fundamentals on the understanding of nature

We have already stated a number of times: The foundation of our insights into nature and our experience consists in the fact that the entire cosmos is comprised of "matter." It is an almost trivial observation that everything, absolutely everything, that takes place in this world is tied to changes in "matter" (structure of substances, spatial position of particles). This also hold true for all processes in living systems.

"Matter" is always tied with mass, this "something" that offers resistance (inertia) to every acceleration, resistance that is stronger the more mass is present. Furthermore, if other masses are nearby, it is even possible to weigh mass, because masses attract one another (gravity). Mass thus has weight which, according to experience, depends on its inner properties and the amount of mass. What are these inner properties?

An understanding of "matter" is the foundation of our understanding of nature and the world!

In this context, I like to mention Giordano Bruno (1548-1600), who was burned at the stake as a heretic, at the behest of the Catholic Church. He said, "There is a spirit in all things, and there is no body so small that it does not contain a part of the divine substance that animates it." For its time, it is a remarkable highlight of human maturity and grandeur, even if we would not phrase it that way anymore.

Otherwise, even today one finds people who take the view that it is not necessary to concern oneself with material questions very much; matter is simply there, and engaging with it is "beneath" us. (This always makes me think of a cow that eats grass, but is interested in neither the grass nor the meadow.)

We have known for a long time that "matter" is made up of various "particles," which are differentiated by their mass and electrical charge, and that they also partly possess a certain "lifespan" until their decay. For the questions posed here, we may assume that it is largely a matter of three particles: electrons, protons and neutrons.

Because masses are given in kilograms (kg), the mass of an electron m_e works out to

$$m_e = 0{,}000000000000000000000000001 \text{ kg} = 10^{-27} \text{ kg.}$$

Furthermore, it possesses a negative elementary charge e of approximately

$$-e = +\, 0{,}0000000000000000001 \text{ coulomb} = 10^{-19} \text{ coulomb.}$$

A coulomb is defined as the amount of charge that, in the presence of an electrical current, flows through a wire in one second when the current's strength equals one ampere.

If one experiments with electrons in certain circumstances, one will further learn that all electrons, in addition to mass and charge, approximately possess the properties of small "spinning tops" and "magnets," whose properties (angular momentum, magnetic moment) are also the same for all electrons and unambiguously definable. These four properties characterize each electron in such a way that electrons cannot be distinguished from one another!

All protons possess a mass that is approximately 1836 times that of an electron, but a positive elementary charge; the same holds true for angular momentum and magnetic moment. The same holds true for neutrons, except that they have no charge.

Protons and neutrons are called nucleons, because they are bound by strongly attractive interactions into atomic nuclei, whose positive charge measures +Ze, if Z is the number of protons in the atom. The properties of the atom are then given by the nucleons. If M is the number of nucleons in a nucleus with a charge of Z, then M – Z is the number of neutrons. The total mass of an atomic nucleus can be approximated by 1836M (atomic weight). This enables us to describe an atom and to make the notes above more precise. An atom is composed of a nucleus (Z, M) with a charge of Z, which is orbited by Z electrons. If the number of electrons is not equal to Z, then the atom can have a positive or negative charge (atomic ion).

This raises the problem of energy; for example, the question of which values of Z and M produce stable (or partially stable) atomic nuclei. In the cosmos, we have only found a finite number of types of atoms in terms of nuclear charge (chemical elements; slightly more than 100).

In this connection we can see, that for a particular Z various values of M that still lead to stable atomic nuclei are commonly possible. We term these atomic nuclei isotopes of the corresponding value of Z, although generally the number of neutrons only varies slightly from Z. For each Z, there are thus generally several stable states of the atomic nucleus in terms of M (stable isotopes). However, we also know of isotopes that decay a certain amount of time after the creation of the atomic nucleus. The time in which half of the amount of the isotope decays (half-life) ranges from fractions of a second to billions of years (radioactivity). Trivially, a nucleus is not stable (thus does not arise), if there is no stable phase and the time of its existence is practically zero; the system is unstable from its very beginning.

As already pointed out above, atoms are tied to the terms of the respective chemical elements by their value of Z. Thus we call an atom with Z = 1 (and one electron) a hydrogen atom (chemical symbol: H). The nucleus can only be made up of one proton, but it is possible for isotopes with one or two neutrons (deuterium or tritium atoms) to exist. H^- denotes a negative hydrogen atomic ion. H^+ stands for a bare proton. Z = 6 is identified with carbon. Uranium has the atomic number of 92.

We stop here the discussion of the atomic nucleus and the stable energy relationships within it, which are only addressed briefly. It is additionally worth noting that in an atom, energy is required to change over to a positive atomic ion (removal of an electron). Correspondingly, energy is set free when a negative ion forms. With these simplified descriptions, we complete the notes about energy relations within an atom.

We may now declare that "matter" is made up of electrons and atomic nuclei, because the energies required for the breakdown of stable atomic nuclei are far greater than the chemical (electronic) relations and possibilities!

Now we come to the decisive question! Because atomic nuclei and electrons carry different charges, they attract one another, and "matter" would thus collapse in on itself, quite contrary to the experiences that atoms exist, and thus their "electron shells" (in relation to the nucleus) must be stable.

We can now present Bohr's atomic model of hydrogen, which was referred to and criticized above, in detail. In the model (planetary model), it was assumed that the electron circled around the nucleus (Z = 1) on precise paths (orbits), which does not reflect experience, as was later recognized. Each orbit was supposed to correspond to a particular energy of the electron, determined by the electrical attraction and the corresponding mechanics. The contradiction lay in the fact that according to electrodynamics, orbiting electrons would have to emit radiation energy, and thus fall into the nucleus; otherwise, the energy differences of the orbits correspond to experience, if one takes quantum theory as the basis. Thus it was assumed that the movements of an electron in each path are radiation-free! As a result of the orbital path, an electron is not distributed spherically around the nucleus, which was not always observed.

Obviously electrons (as well as protons and neutrons) are particles of a new and different sort, because in the larger amounts of material that occur in daily life, we do not observe the phenomenon of indifferentiability. Precisely the differentiability of objects is a fundamental fact, without which there would be mistakes and confusion.

I have always been surprised that the unsettling fact of the indifferentiability of electrons, an important material particle, draws so little attention and is generally accepted, particularly in science but among the general public as well. Precisely this fact is the reason – one can show it theoretically – that for a particular Z completely similar atoms exist that differ only in the energy states of their electron systems. Otherwise, for a given Z there would be atoms that behaved completely differently, making the relations in chemical bonding confusing.

If we observe more atoms, forces between the atoms emerge, which derive from the electrons and which overlay (sum up) the forces of repulsion between the positive atomic nuclei (coulomb repulsion). These forces are, with a few exceptions (noble gases), attractive, so that collections of atoms (molecules, crystals, metals) can form at all! These forces have a fairly complex construction, which we do not need to examine in detail. Because of the indifferentiability of electrons, these processes depend – in addition to the number of electrons and their total energies – only on the electrons' angular momentum and magnetic moment.

It remained to be shown that the construction of the atomic shells in the atoms in question fundamentally determined the character of these forces (atomic interactions). Because the forces of repulsion between the nuclei increase without bound as the distance between the nuclei decreases, while the interactions between electron shells, by contrast, always remain finite as the distances between the nuclei shrinks, an equilibrium is reached in which the repulsion of the nuclei and the attraction of the shells balance one another. This produces "distances" between the atoms in "matter" – or at least that is how one usually reads it in the literature.

In fact, we have just presented the relationships in a simplified fashion and drawn up a model to make the situation comprehensible, the more so as it seems to encompass the essentials. Reality, however, presents things differently. There are no set distances between the atoms, and thus the atoms themselves are also not tied to a set position (average distances)! Matter is in fact something quite different; for that reason, we have occasionally set the word "matter" in quotation marks, to make that clear. Wave mechanics was the first approach to deliver a correct description of relationships within matter, by introducing the terms of probability into our understanding of nature. In this context, one often hears the expression "zero-point vibration," which does not apply either, because it is not a question of vibrations!

The forces of chemical bonds, and the bonding energies that are tied to them show a form of "saturation." To present this set of facts, we observe the simple case of bonding between two H^- atoms, which can form an H_2 molecule. If an additional hydrogen atom approaches the H_2, this is repulsed; no H_3 is formed. If, however, an H^+, that is a naked hydrogen nucleus (proton), approaches H_2, then H_3^+ can be formed, a positive molecular ion of three hydrogen atoms. By contrast H_3^-, that is, three protons and four electrons, is not to be found. Under certain conditions H_3^+ decays as follows:

$$H_3^+ \rightarrow H_2 + H^+,$$

although

$$H_3^+ \rightarrow H_2^+ + H$$

can also be observed. The H_2^+ is a hydrogen molecular ion with two protons and one electron, which "manages" the bond between the two protons. One would also expect that the bond distance in H_2^+ would be greater than in H_2 (H_2^+ and H_2 in ground state), in the absence of special conditions.

Finally, we should note again that the two partners in a bond not only maintain the bond distance, but also, if less commonly, greater or lesser distances from one another. (This is often interpreted as oscillation in the equilibrium distance, which is what one finds almost without exception in the literature, but which, as mentioned, is incorrect!) It is again a matter of a model.

Wave mechanics describes the situation quite differently, and in a way that agrees with all of the experiences, which we will discuss later, along with the implications that follow.

Chapter 6 – Matter generally observed[3]

The detailed treatment of relationships in the hydrogen atom facilitates a transition to all atoms. The bonding mechanisms between two atoms, A and B, are of the same kind; a mean bond distance arises at which A and B most commonly remain, although repulsions can also occur. Various electron states produce various median bond distances!

With three of more atoms, the interaction – determined by electron states – lead to varying molecular spatial structures, although saturation effects again play a role.

With water (H_2O), the three atoms H, H and O (oxygen) form an isosceles triangle, in which only the two O-H bonds arise, which then repel each other (no bond forms between the two hydrogen atoms). Here as well, no third hydrogen atom is bonded.

By contrast, H_2O^+ is known. H_2O has ten electrons, $(H_3O)^+$ also has ten electrons; it is also constructed with the oxygen in the middle, with the three O-H bonds at angles of 120 degrees. O_3 (ozone) presents an isosceles triangle, and CO_2 (carbon dioxide) is elongated (linear), with the carbon atom C in the middle. One final example: In benzene (C_6H_6), the carbon atoms are in the middle of a level ring with the same median distance to their neighbors. A hydrogen atom is attached to each carbon atom, while the system stays planar.

"Bending" C_6H_6 requires energy! Benzene contains 42 electrons, which are substantially divided, in the middle, and around the nuclei, and between them, although six electrons can be found around each carbon atom in the middle. Benzene is thus saturated. It does not take much imagination to envision what an uncountable multitude of possibilities exists as the number of atomic nuclei and electrons increases, although here we have only shown the most fundamental aspects of molecule formation. One should consider that there more than 100 elements in the cosmos, and the number of electrons available for the

3 In this context, see also Preuß, H., Atome und Moleküle als Bausteine der Materie, Salle: Sauerländer 1982.

interactions of atoms and molecules is practically unlimited. Likewise for the formation of molecules, and the resulting generation of gas, liquid and solid materials, there are practically no limits. In this respect, we are thinking of crystals and metals, and also of wood, glass or rubber.

Matter can thus take on all of the characteristics that we know from our experience: plants, trees, animals; humans, including their brains, are also made up of matter, of electrons and atomic nuclei. Matter can, for example, have color, emit odors, show radiance, conduct electrical currents, be magnetic, and much more.

All of these experiences with matter, and also their behavior under external influence, of whatever kind, can always be linked back toward the behavior and reactions of systems comprising electrons and nuclei, as we now know! In these actions, we can also presume that in all of these processes individual atomic nuclei remain unchanged, with the exception, noted above, of (natural) radioactivity.

Even under extreme conditions, as can be found, for example, in certain cosmic locations, in which atomic nuclei change or decompose and other particles emerge, our assertion above retains its full validity, because matter as such remains conserved!

That makes these questions even more fundamental: What, really, are electrons and atomic nuclei? Is our previous conception of matter still valid?

Chapter 7 – Some thoughts on humans and matter

The assertions just made – which are fully justified and show no exceptions – appear to be trivial. Why in fact should the entire cosmos not be comprised entirely of matter, which in turn is entirely "built up" of electrons and atomic nuclei? But some readers may already be having doubts: is it really true that everything is made up of particles of matter? Can even thoughts, feeling and consciousness really arise from just the structures of matter? Without influence from outside? That is, from outside of matter! Does mind have a material nature? Is there no creator – a god – who intervenes and makes everything possible?

Whoever holds fast to "considering to be real" has nothing on these doubts and thoughts. That view of the world is determined by dogma, and the behavior is firmly fixed.

But in light of our knowledge about matter, which we will soon examine more closely, doubts must be allowed, because even just the insights presented above about the composition of matter should inspire reflection! Because if one becomes more consciously aware of it, one realizes that these conclusions pertain to every person, and can be conceived as an almost sensational insight! At least for people who want to consider the problems of the universe impartially. That makes it the more disappointing to realize that the present situation goes completely unrecognized by almost everyone!

That has many reasons, which can be found in the structure of society. Undoubtedly, sufficient basic knowledge of the natural sciences is one of the things that is not taught in the world's schools in such a way that the students can put it to use later and be able to put it in context with other questions, which is actually the beginning of a true education. Thus examination of matter falls by the wayside, and emphasis is placed elsewhere. But teachers themselves lack deep understanding (setting aside some exceptions, as they are a minority that has no effect on the public). The media do the rest, and nothing is done from the political side either.

Technology is worshiped everywhere, in which science is seen as the source of technology, exploitation of nature, without seeing into it, is all "at best" suitable as entertainment.

This type of criticism of the current situation in society, in regard to its behavior toward matter, must be dealt with from various sides. Because a discussion about the implications of the insights that are available today can only proceed if it is clear from the beginning for which reasons and out of which situation human society has failed to draw these conclusions, and mostly has not been allowed to draw these conclusions.

Many so-called experts, who actually know the facts, or at least should have heard of them, either close their eyes to them, or assess the situation differently because the wrong notions are always repeated in the media and in the schools.

Furthermore, in the humanities there is an understandably widespread aversion to an intense and deep engagement with natural science, because the scientific way of thinking itself contradicts the general philosophical way of thinking (again, overlooking some exceptions).

The connection between observation and insight is completely undervalued. Thus it could come to pass that in recent centuries an unbelievably vast, intellectual, partially original philosophical literature has been created, which of course has earned respect and belongs to our culture, but which lacks clear and conclusive insights – precisely an understanding – about our world. That is also the source of the aversion to mathematics, whose deeper meaning is barely recognized! Even an engagement with matter is seldom found. That makes it the more astonishing that in the fifth century before the common era Leucippus of Miletus, along with his pupil Democritus, had already developed the idea that matter was made up of extremely small, indivisible particles – atoms – although the origin of this word dates back to the Hellenistic period. Leucippus even distinguished among various types of these "indivisible" particles, although these concepts were borrowed from experiences with macroscopic matter. A very mature philosophical achievement, which deserves admiration, but what became of this idea? Centuries went by, in which matter was forgotten.

By contrast, once certain "implications" are drawn from the humanities, implications that affected social topics or ethics, these developments were destined for intellectual failure. I mean this because they lacked a foundation in the natural sciences, even if they were sometimes presented in a pseudo-scientific fashion.

The same holds true for the realm of belief. In the field of "considering to be true," matter is completely excluded – or at least considered in such a way that it is not necessary to think about its structure, although its variety cannot be overlooked.

Even today, now that the existence of electrons and atomic nuclei has been recognized, this fact is considered unremarkable, although it has cosmic validity. We speak constantly about the world, use matter, but knowledge its inner structure and the implications thereof do not matter to us. Even the consequences of "using matter," which has increasing disadvantages for humanity (e.g., weapons, environmental damage), are in the end accepted because the interests are completely askew.

If we look at a beautiful crystal, we seldom ask what it is made up of, but instead about its price, and many who do are at the same time thinking whether they could obtain it somewhere else for less. That is the link between material understanding and capitalism! The "price" in the most general sense blocks our view of what is really natural, of deeper understanding, of insight into what leads more closely to questions of meaning.

I hope, with these considerations that I have presented in several variations with repetition, to have shown how close the relationship between our understanding of matter and our relationship to the world, and thus our fellow humans, really is, which also raises questions of morality and ethics.

If one wants to understand the situation of humanity, one cannot leave out the relationship to matter! Despite all philosophies and religions, it does not look good for the creation of a universally valid and acceptable system of morals and ethics!

We should honestly admit that human development over the course of time has brought more disorder than true progress, which should consist of increasing the number of people who live reasonably, contentedly and morally. The opposite is true, even if we allow ourselves to be convinced of the meaning of progress that is propagated by interest groups. Disloyalty extortion, wars, murder and torture are increasingly a part of daily life; follies and primitivity (strengthened and supported by the media), lack of judgment and lack of insight are increasing, and with them fanaticism and widespread arrogance.

It will not be possible to develop a global morality and ethics, because several claims will be made that pretensions of totality, but which contradict one another – that will lead to conflict!

The existence of a capitalist society, which now encompasses the entire globe, stimulates the relations above without limits, because has begun to transform everything – people, humanities, philosophy, and finally religions as well – into goods, into an opportunity for capital to develop its power. Out of all this results a general disruption of the human process of maturation (long-lasting adolescence, often childhood).

Let us put it more precisely, to be prepared for later discussions: If a certain part of nature, of matter, which possesses sufficient complexity, changes into a money society, and thus as a result captures the rest of nature for its purposes, exploits it to concentrate capital, and does this with all of its abilities; if it attempts to convince others who do not (or cannot) join in that all of this is necessary and good, in the end it is acting against all of humanity and is obstructing the future, the survival of all. This conclusion will be clearer later.

It is unimaginable for people – as indeed it must be in the current situation – that deep scientific understanding of the cosmos leads to a code of behavior (or perhaps indeed must lead to one!). But precisely that is the case, as we shall show.

Chapter 8 – The path to understanding matter

At the end of the 19th century, a problem of physics became increasingly prominent, one that stubbornly resisted a solution – being understood. It was a question of the behavior of matter in its key meaning for our experiences.

It was a question of research into the emission of electromagnetic waves (radiation) from matter as a function of the matter's temperature T, in which the nature of electromagnetic radiation had already been explained: an electromagnetic wave with frequency ν (oscillations per second) which propagates at the speed of light c (300,000 km/sec). Electromagnetic radiation includes radio waves, infrared radiation (heat radiation), light, ultraviolet radiation and x-rays.

More precisely: A heated body (such as a piece of metal) normally first emits infrared radiation, which is invisible to the human eye. However, if the temperature is further increased, one observes, as is well known, a faint glow that, with additional temperature increases, changes into bright light (higher ν). Eventually the body will shine light blue (increasing ν); it begins to emit ultraviolet radiation before further temperature increases cause it to melt and then later to vaporize.

All parts of matter behave in the same way, according to their composition. This is thus a fundamental behavior of solid matter. For every temperature T, there is a certain distribution of radiated energy across the frequencies.

To eliminate the varying compositions of matter, measurements were oriented on a so-called "black body," a piece of matter that completely absorbs electro-magnetic radiation of all wavelengths λ (all ν) $\nu = c/\lambda$.

The property of complete absorption is the point of reference and the pre-condition for measuring the release of radiation as a function of temperature, as described above. Carbon black comes very close to meeting these requirements, but it is not very well suited for measurements.

Investigators discovered that a hollow body with walls that were impermeable to electro-magnetic radiation but with a small opening completely fit this ideal. The small opening behaved like an idealized black body when the hollow body was brought to a constant temperature T. This opening emitted the radiation as discussed above.

The question then became how the complete radiated energy was distributed among the various frequencies, depending on the corresponding temperature. This distribution thus depends on T, as explained above. For each temperature of the hollow body it is possible to measure at each frequency ν a particular amount of energy ε, such that overall a curve $\varepsilon = \varepsilon(\nu)$ is generated, which still depends on the temperature.

Until 1900 it had not been possible, with the help of theoretical considerations, to compute such a "radiation curve," which had to mean that the theory that had been developed up until that time – including thermodynamics, electrodynamics and Newtonian mechanics – was evidently not up to the task of describing this process correctly; that is, in way that was consistent with experience. At the time, this was a considerable surprise, because no one had expected that this apparently simple process would require an extension of theory! Apparently, matter was not yet completely understood; or in other words, the behavior of electrons and atomic nuclei needed to be better comprehended. More insight was needed! If this succeeded, then it could be hoped that the behavior of electrons and atomic nuclei could be understood, and thus of matter in the entire universe!

For a philosopher it may seem almost absurd that such conclusions can be drawn from a "simple" process of emissions, conclusions that go far beyond the philosophical concerns, if indeed there is any interest in these possibilities. But it is precisely here that the key difference between philosophy and natural science lies, although precisely the former (as with religions) makes assertions about the universe, and considers itself responsible for such. Every person should consider these issues and take them seriously – as they matter to everyone directly, as we will show more precisely later.

The solution to the problem above, which Max Planck ultimately proposed as part of his lecture to the Prussian Academy of Science on

14 December 1900, and which proved to be fully correct as it easily produced the correct radiation curves, was the beginning of quantum theory! Never before had similar thoughts and conjectures been discussed.

Today it is interesting and informative to read how developments proceeded, and in particular how surprised Max Planck himself was that he had to introduce an assumption that led to the derivation of the correct radiation formula, an assumption that he believed would be temporary, only necessary until additional findings had been made. For many decades previously, many physicists had tried unsuccessfully to find the solution. One could say that Max Planck was virtually "forced" to find the truth, while he was looking for it!

At any rate, the correct radiation formula $\varepsilon(\nu)$ could now be found via consideration and calculation, if one maintained the theory up to that time, but added an additional important assumption; namely that the energy E emitted by the atoms and molecules (not ε) of the material in question is related to the emitted frequency ν according to the relationship

$$E = \text{constant} \times \nu \tag{2}$$

That the existing theory continued to be valid and only the addition of equation (2) was required showed that the outlines of an extended theory had been found, particularly, and importantly, because Max Planck's idea was repeatedly confirmed by the understanding of additional experiences. The same constant in (2) was always yielded, such that experiences were captured correctly.

$$\text{Constant} = h = 6{,}625 \times 10^{-34}\ \text{J x sec.} \tag{2a}$$

(E and ε should not be confused with one another, because ε as a function of ν refers to the energy that the material emits, whereas E refers to the energy emitted by the individual atoms and molecules of which the material is comprised, and atoms and molecules are comprised of electrons and atomic nuclei.) One gains insight into and understanding of nature if one extends the existing theory with relationship (2).

Planck referred to the constant in (2) with h, and it has remained so through today, thus:

$$E = h\nu \tag{3}$$

The designation to the right of h in (2a) names the dimensions of the quantities. Energy is measured in joules (J), so that h in (2a) refers to energy times time, which corresponds to an action. The constant h is thus often also referred to as Planck's quantum of action, because according to (3) clearly for every ν a certain amount of energy, a quantum of energy, is correlated.

The constant h is a very small amount, and occurs in all processes in the cosmos! It has proved to be a constant of nature, like e or m_e.

Equation (3) did in fact enable the explanation (by Einstein) of the photoelectric effect, which was also known at the time. This effect consists of electromagnetic radiation with the frequency ν in a vacuum falling on a metal plate, and beginning with a certain ν electrons leave the plate. The distribution of their speeds, however, is such that the maximum speed of an electron leaving the plate does not depend on the intensity of the radiation, but instead on the frequency of the light beamed at it!

The explanation of this set of facts (Einstein 1905) succeeds with the help of (3) in that its conclusion must be made more precise. This must namely be interpreted to mean that every emission is comprised of certain "portions of energy" $h\nu$, which were later known as photons, so that light must be considered as quantified.

Because, however, of Einstein's well-known equation

$$E = mc^2 \tag{4}$$

Energy E divided by the speed of light c is equivalent to mass m, it can be inferred that photons possess the mass m_{Ph}, which one obtains by combining (3) with (4).

$$h\nu = m_{Ph}\, c^2. \tag{5}$$

From (5), one obtains

$$m_{Ph} = \frac{h\nu}{c^2} = \frac{h}{c\lambda}, \tag{6}$$

if one applies the well-known equation from optics (see above)

$$\lambda\nu = c \qquad (7)$$

in which λ is the wavelength of the electromagnetic radiation.

Now we can see a bit more clearly (model): The photons "knock" the electrons out of the material (metal plate), and they must have a particular energy for that to succeed because the electrons in the material are attracted by the atomic nuclei. The attractive force must first be overcome by the "hit." It thus follows that:

The smaller the wavelength, the greater the frequency (see (7)) and thus the greater the mass of the photon (see (6))!

From this perspective, light is a "particle emission" comprised of photons! But that is only one aspect of light that undoubtedly applies in this picture.

If we speak of a "hit" between a photon and an electron, we are using one model of the concepts. According to wave mechanics – which developed from quantum theory – this is not permissible! We cannot make any precise assertions at all about this process; nevertheless, the statements made above hold true, if one does without certain details, as we will later recognize more clearly. In 1927 Compton was able to completely confirm the basic process with ingenious experiments!

But the development was not yet complete! There was now a new problem, which was related to light, because light (electromagnetic radiation) appears to us as a wave, but also as a particle (photon). One tends to see that as an insurmountable contradiction, and even today, in the early years of the 21st century, some physicists still see it that way. Thus it is of course no surprise that schools (and universities) and the media speak of a wave-particle dualism if the topic arises; but this dualism does not exist!

Without much consideration, two typical experiments (experiences) are often performed that "obviously" contradict one another: the photoelectric effect, which we discussed above (light as a stream of photons) and diffraction experiments with light (wave emissions). In a diffraction experiment wave fields are overlaid on one another, so that the waves' peaks and troughs cancel each other out. Because of this typical wave effect, when light passes through a narrow opening onto a screen behind it, numerous images of the original opening appear on the screen.

This phenomenon can apparently only be made understandable with the wave conceptualization of light!

Advancing experimental techniques demonstrated that electrons, nucleons, atomic nuclei and even molecules showed diffraction effects. Obviously this approach had led to the discovery of a special property of matter, one that first appeared as a complete contradiction. On the other hand, it was soon clear that its solution would reveal a completely different understanding of matter, and that is why we will examine the circumstances more closely.

To do that, we start from another side. J. C. Maxwell developed the theory of electrodynamics, or magnetism, in 1864 and was the first to recognize the existence of electromagnetic waves, which were soon (1886) demonstrated experimentally by H. Hertz, confirming the predictions of Maxwell's theory.

At this point, we should be clear that the entire difficulty arose from Planck's assumption and thus from the conception of the photon as an expression of the electromagnetic field as described by Maxwell. The photon has (according to equation (3)) the energy hν. It is interesting to note that at first no wave field (emission) could be determined for the other particles, whose "wavelengths" would explain the observed diffraction effects.

Fundamentally speaking, a generalized theory would have to be found, one that goes beyond quantum theory and includes Maxwell's theory. We thus determine again more generally that clearly every particle of matter must be correlated to a wave field, such that by passing through an opening (as a wave) the wavelength "λ" occurs, which makes the diffraction image optically understandable.

A particle model does not allow for such considerations, particularly because it has been shown that electrons, nucleons and atomic nuclei can possess not only mass and charge but also spin and magnetic moment! Photons also possess spin, as was soon demonstrated.

In this context, an important experience was that the particle stream that passed through the gap must be comprised of particles with the same speed (always the case for photons), so that the diffraction image appears; otherwise indistinct images would appear, blurred into unrecognizability.

In considering these contradictory phenomena, we must ask ourselves where we have gone wrong in developing our views about matter and about the world in general over the course of the millennia; because without "material knowledge," no view of the world, no understanding of the world can be developed or built up, not least because matter throughout the universe is comprised of electrons and atomic nuclei. That also holds true for particles that can be formed under extreme conditions in the cosmos.

At any rate, it became clear that matter can not be what it is so commonly considered: something dead, completely lifeless, the opposite of spirit and soul, and certainly not a topic for feelings. Matter pulls us down into the realm of the soulless, the anti-spiritual. No – this much is already clear: Matter must be more if it presents itself to us with such extreme contradictions.

At the same time, we ourselves are also made of matter, the entire world is made of it, and if it is a matter of questions of chemistry, biochemistry, pharmacology, ultimately and comprehensively of all sciences, philosophies and religions, of all phenomena and experiences, then our cosmos (and we with it) is in fact comprised of photons, electrons and atomic nuclei; that is the well-founded scientific finding, and no claims of "believing to be true" – for whatever reason – can change it. Whatever we touch, see or smell, we always make contact via these three "particles," which do not seem to want to fit into our understanding of particles, which compels us to consider that all processes in us and around us have to do with matter. – What does the solution really look like?

Let us first note that in general the momentum p of a particle is defined by

$$p = mv \tag{8}$$

in which v is the velocity and m is the mass of any given particle. The momentum should not be confused with the particle's energy, which is given by

$$E = \frac{1}{2} mv^2 \qquad (9)$$

and describes the energy of motion (kinetic energy). These equations are derived, as is well known, from classical mechanics.

Now we multiply equation (6) on both sides by the speed of light c, so that the results of the equation remain unchanged, and thus come to the relationship

$$m_{Ph} c = \frac{h}{\lambda}. \qquad (10)$$

We can now recognize that on the left side of the of equation (10) the momentum of the photon has appeared, because only this particle moves at the speed of light, thus (10) becomes

$$p_{Ph} = m_{Ph} c = \frac{h}{\lambda}. \qquad (11)$$

At this point, we should remember that we have obtained (11) as a result of bringing together Planck's assumption (3) and Einstein's conclusion about the equivalence of mass and energy.

In 1924, De Broglie proposed extending equation (11) and using, in place of the photon's momentum p_{Ph}, a more generalized momentum as described in (8) and then solving the equation for λ (a "hypothesis" according to De Broglie, which later proved to be a theory).

$$\lambda = \frac{h}{mv} = \frac{h}{p}. \qquad (12)$$

This equation constitutes a sensation! It links a wavelength (λ) with a characteristic of a particle ($p = mv$). And this is made possible by Planck's constant, h! It could not be more clearly stated how important h is for the understanding of nature.

It became clear that this line of thinking had been correct, because the wavelengths that were calculated for particular speeds and masses of the particles correspond precisely to those that are necessary to calculate the diffraction pattern – and also to predict it!

Has the contradiction in our experiences with matter been truly resolved already? We have in fact mathematically tied together the two conceptions of the wave view and the particle view, justifiably and reasonably, but we do not yet know what this equation means. Why was it possible to tie the two contradictory conceptions together in this manner in a single equation that manifestly describes experience correctly?

This is a good example for showing the role of mathematics in theory and in hypotheses. To put equations (1) through (12) into words would have required an enormous effort, and at the same time one would not have attained the same clarity about the observations! I mean mathematics and not computation, which are two entirely different things!

In mathematical representation, symbols are meaningfully tied to quantities, within the framework of theories, as for example m with mass, velocity with v and wavelength with λ in (12). "Meaningfully" means that the connection conforms with experience, that is accumulated with these quantities (measured quantities). Computation, by contrast, presumes numeric values for m, v and λ (or ν), which complete the equation (12) and make is possible to verify the measurements numerically.

The mathematical presentation of the relationships in nature – that is, the formulas – has an additional advantage that nearly characterizes them. We survey the relationships and recognize from (12) that the wavelengths are conserved if we keep a particle's momentum constant. We recognize that particles with greater mass must move more slowly so that λ remains constant. On the other hand, equation (12) "reveals" that if the constancy of momentum is lifted, greater masses at the same speed will lead to shorter wavelengths. Finally, it is evident that λ can only exist if Planck's constant is taken into account.

Demonstrating all of this without formulas is laborious and very prone to misunderstandings and misinterpretations. From this perspective, mathematics is in a certain sense an "economy of thought"! Lack of mathematics bars the way to a deeper understanding of nature.

The success of the natural sciences, and of physics in particular, is in the end based on using mathematics so perfectly to understand nature. The same also holds true, in terms of precision and insight into function, for technology, which is largely based upon the natural sciences.

Chapter 9 – Wave mechanics

In 1925, Werner Heisenberg stated a mathematical relationship that went much further than (12), without making it invalid – quite the opposite, in fact, it incorporated it. Heisenberg's proposition pertained to the location and the momentum of every particle at a given time. If we introduce the Cartesian coordinates x, y and z, all at right angles to each other, then it is a matter of the location x, y, z and the momentum p_x, p_y, p_z (along the x, y and z axes) of the particular particle. Heisenberg's formula, known as the uncertainty principle, states a relationship between uncertainties in location (Δx, Δy, Δz) and the momentum (Δp_x, Δp_y, Δp_z), and does it in the form of inequalities (a greater than or equal to b means that a >= b).

$$\begin{aligned} &\Delta p_x \Delta x \geq \hbar \\ &\Delta p_y \Delta y \geq \hbar; \qquad \hbar = \frac{h}{2\pi}. \\ &\Delta p_z \Delta z \geq \hbar \end{aligned} \tag{13}$$

It must however be emphasized – and that is the mind-boggling conclusion – that these inaccuracies exist as a matter of principle, that they are a law of nature, as we have recognized! These inaccuracies (or blurring) are always present, if we want to measure the momentum and location of a particle at the same time.

The extent of this conclusion, this law of nature, can hardly be overlooked. And its consequences are devastating for the old conceptions of natural science, including those of philosophy and religions, if these presume that all parts of matter have precise locations and speeds at any arbitrary time, even if they possess an extremely small mass!

Equation (13) reveals that this is never the case, even if the effect is not so significant for larger masses; the uncertainty never disappears. The inaccuracies (or blurring) are not an artifact produced by the instruments, they are a property of matter!

The shrinking influence of the uncertainty as mass m increases can easily be recognized if one uses equation (8) to take the mass out of the momentum on the left-hand side of (13), so that the right side contains $\hbar$ divided by m, and thus the right side of (13) becomes smaller as mass increases. The product of Δx and Δp_x will thus be made smaller as well (as is also true for y and z).

For most of us, this is all apparently immaterial and irrelevant, but I would warn against holding this opinion, because one should not forget that matter is comprised of electrons and atomic nuclei, which are particularly affected by the uncertainty! Thus with (13) matter is encompassed and described.

At almost the same time, 1926, Erwin Schrödinger drew up the partial differential equation that bears his name, and which present a comprehensive theory of matter that encompassed all of the relationships, without exception, that had been recognized up to that time! It soon turned out that a theory Heisenberg developed at the same time, known as matrix mechanics and which proceeded from differing assumptions in both mathematics and overall thought, was fully equivalent to Schrödinger's version, which is truly remarkable and speaks in the theory's favor. Because the Schrödinger equation is simpler to work with, it is used in the present day almost without exception.

In the formation of his equation, Schrödinger drew heavily, among others, from λ according to De Broglie, and described an oscillation process in matter. This led him to his equation, which is, in a sense, an oscillation equation. The time-dependent Schrödinger equation, and the additional conditions tied to it, describes the behavior of matter – electrons and atomic nuclei – completely, including all possible temporal processes!

That is the foundation of wave mechanics (quantum mechanics)! Any subsequent generalized theory must include wave mechanics because it has been experimentally demonstrated and its mathematical version can be formulated without contradiction. The foundation is, as we have demonstrated above, completed without any exceptions. Any implications are thus not just a matter of being believed to be true. They are valid throughout the cosmos, as has been demonstrated time and again. We will shortly be examining that aspect more closely.

In this context, it is important to keep in mind that there is no alternative to wave mechanics, which is what distinguishes the "truth" of natural sciences, because any alternative would have to describe the same experiences! The formalization of such an alternative would be, as the term has it, isomorphic.

Before we examine the equation, we will need to depart from the usual confines of the CGS-system and switch over to the system of atomic units. That will make the mathematical presentation of wave mechanics formally independent of the natural constants, and also simpler because all masses and charges will be measured in terms of electron masses and charges (m_e, e). The unit of force is $\hbar$. This produces the other measurements, such as length, speed and energy. Accordingly, the time-dependent Schrödinger equation is

$$H\Psi = i\frac{\partial \Psi}{\partial t};\ \left(i = \sqrt{-1}\right). \tag{14}$$

This mathematical relationship describes the behavior of an arbitrary number of electrons (n) and atomic nuclei (N), in that the Hamilton operator H contains information about the system which, according to (14), should be investigated. However, H also presents a computational instruction (operator), which should mathematically occur with the function at its right, which also depends on n and N. This computational instruction comprises the multiplication with certain mathematical expressions and of differentiation on the location coordinates of the n+N particles, thus according to 3 (n+N) coordinates.

Thus H is always known! The charges of the N atomic nuclei are also incorporated in H. In equation (14) another function, Ψ, appears, which was unknown until the formulation of wave mechanics! It likewise depends on all of the particle coordinates, as well as on the time t, although we should also mention that Ψ also contains additional characteristics of the particles, such as their spin (angular momentum). Ψ shows the oscillation characteristics as a function of time, and for that reason is known as the *wave function*. Once H is known (as a system), the theory's task is to solve the Schrödinger equation, that is, to find a Ψ that balances equation (14) when it is plugged in.

Schrödinger's equation (14) contains the so-called imaginary unit, i, such that Ψ also contains this unit. We thus say that Ψ is a complex function. If in Ψ the unit i is replaced by –i, we write Ψ^*. These mathematical operations were known in the realm of complex numbers before the development of wave mechanics. The usually known numbers are called real numbers.

According to a mathematical identity, it is always true that

$$\Psi^*\Psi = \text{real magnitude}, \tag{15}$$

which makes the introduction of Ψ^* understandable, because only what is "real" can be measured!

We must address these apparently abstract facts here because otherwise it would not be possible to present the subsequent consequences, which are our ultimate interest, sufficiently convincingly. I am touching only on the most important elements, and simplifying, as long as it is not at the cost of the scientific nature of the inquiry, which would limit the importance of the conclusions! Concerning Ψ as the solution of the Schrödinger equation, we differentiate between approximate solutions, super-tilde-Ψ, which do not solve (14) completely from those which we know to be approximately in accordance with sub-tilde-Ψ, particularly in certain characteristic properties (mathematical behavior). The mathematical symbol for "approximately" is $\approx$, so that we may write

$$\tilde{\Psi} \approx \Psi. \tag{16}$$

In a calculation of error, the behavior of $\tilde{\Psi} - \Psi = \Delta\Psi$ is discussed, which can be very important if, for certain reasons, a calculation of Ψ is not (yet) possible.

Ψ (and equation 15) clearly describe nature – matter – and can be set in full relation to it, in that all possible experiences with matter can be brought together in this fashion with mathematical symbols and expressions, which is of course the key characteristic of a theory, as we demonstrated more generally above. That, however, is getting ahead of ourselves a little bit.

The last and decisive step is still missing! For a time it was unclear what should be understood with Ψ, and thus with equation 15. On the other hand, it was realized that the equation described matter correctly because all of the stationary states of electrons and atomic nuclei, which had a particular – and temporally unlimited – energy state, were found to be in accordance with experience, and were also predicted where no previously measured values were available! Furthermore, in time-dependent processes scientists increasingly concluded that the time-dependent Schrödinger equation correctly described natural processes.

It is typical for a good and relevant theory that the correct formal presentation also leads to quantities that are new and which must first be interpreted and incorporated (defined). In the case of Ψ, it is clearly a key quantity, whose mathematical form (solution) is certain. At the beginning, however, researchers were uncertain how it should relate to experience.

Soon, though, Max Born offered the decisive interpretation of Ψ, whose consequences were hardly foreseeable, which created a wholly new way of looking at the world and dissolved all of the previous contradictions. Furthermore, it was the first time that a solution in this particular form had been seen in physics. It also resolved the wave/particle contradiction: there is no duality, and its disappearance opened the way to a comprehensive understanding of the world.

Ψ itself remains a complex function (dependent on i), and is for that reason not a quantity to be measured, what a physicist would call an "observable."

Born proposed considering equation (15), which is real, as the probability density of n + N particles in a space.

More precisely: $\Psi^*\Psi$ depends on all of the particles' spatial coordinates. That means that $\Psi^*\Psi$ is a 3(n+N)-dimensional function, that has a particular value for every spatial point within this "hyperspace." This "point" corresponds to a particular configuration of the n + N particles in our three-dimensional space. Then we construct n + N miniature cubes $\Delta\tau_j$, one for each particle (j – 1, ..., n + N).

Thus

$$\Delta\tau_j = \Delta x_j \, \Delta y_j \, \Delta z_j \tag{17}$$

with the edge lengths given by the right side of equation (17) and presume that each cube contains a particle, which means that the three spatial coordinates of the j-th particle lie within $\Delta\tau_j$.

Born's proposition is then written in the form

$$\Delta W = \Psi^{*}\Psi \, \Delta\tau_1 \, \Delta\tau_2 \, \dots \, \Delta\tau_{n+N}, \tag{18}$$

in which ΔW represents the probability that the (n + N) particles, as discussed, can be found in the cubes. The spatial coordinates of the (n + N) particles are also inscribed in $\Psi^*\Psi$ and provide a specific value for (15). That is then valid for every time t, since Ψ depends on t, and thus ΔW contains time, although this does not necessarily have to be the case (for example, with stationary states, as described above).

In this process, we presume that the cubes are selected to be sufficiently small that ΔW only changes a very little if we change the coordinates within the cube. The spatial situation of the n + N cubes can be freely determined, so that we are able study the spatial structure. That is crucial information about the particle system! The results are always in agreement with experience! They thus allow for predictions as well as subsequent confirmations. Equation (18) thus includes the conclusions of the principles of indeterminacy.

Because Ψ also includes the individual spins of the particles (called spin coordinates), ΔW is also dependent on these values. In this relation, it must be kept in mind that the spins can only take on specific spatial values (quantization of spatial directions), but which we do not wish to explore in greater detail. This assumption, too, is in accordance with experience.

Finally, it should be noted that one relation should be added to the indeterminacy principles (13), which makes an assertion about energy and time:

$$\Delta E \Delta t \geq \hbar. \tag{19}$$

Energy and time are also encompassed in an uncertainty principle. There can be no function E = E(t) that would describe the process of radiation! We do not, however, wish to pursue this thought more deeply.

Chapter 10 – Extensions, Pauli principle, summaries and examples

We have identified the fundamental elements of the wave mechanical theory of matter, although it is crucial to point out that an understanding of matter, and thus of all of nature, is only possible with the assistance of a certain degree of mathematical knowledge. Otherwise – and this must be vigorously asserted – human understanding of nature, that is, insight into the relations of the world, is closed off, or at the very least made much more difficult. Mathematical thinking, what we have above called an economics of thought, is accessible to anyone by nature, and thus belongs to education (see above); otherwise, a person is at the mercy of things believed to be true, to belief, as an approach to life. As has been previously shown, the necessary mathematical background is relatively small and can be kept to a limited amount, so that the acquisition of this knowledge is easily possible.

We will return to this crucial aspect repeatedly, because there is no doubt about it, and human history shows that necessary insight – and not merely believing things to be true – into the environment is critical to survival, all the more so as technology and science rise, and the influence of the media prevails!

A belief can, in principle, be changed, as there are many religions, each of which strives to be absolute. Knowledge from natural science cannot be changed, because it is secured by scientific and theoretical experience (structure), and thus has no alternative. We will soon be able to demonstrate that this knowledge reaches even deeper and further! In order to reach this conclusion more readily, we have repeated some of the propositions related to this context in other words.

There is, however, no doubt that the other approaches also have their historic importance. Without belief, without something that people could hold on to and believe to be true, they would hardly have been able to leave their archaic existence. In the due course of time, belief can also lead to a certain level of maturity; likewise, there is no doubt that philosophy belongs to all persons who think deeply about the world and themselves, and that always defends the humanistic disciplines.

But almost everything has its time, especially if it includes societal parameters. Thus increasing population, late-stage capitalism that is increasingly wild and ruthless, together with increasingly puerile media and societal influences, have led to a situation in which religions are strongly involved with conflicts (as in earlier eras) and, in fact, are very often their cause.

That is because people who believe something to be true must defend and demonstrate this point of view to "unbelievers;" and thus in many cases it is not so very far from fanaticism. Only a cosmic ethics can dissipate these tensions, an ethics that we will come to in due course, as it is closely related to the scientific results whose origins we are currently developing.

A cosmic code of behavior would include us in the society of all sufficiently complex systems in the universe, which would, at the least, simplify any possible contacts in the future. Relations would be defined *a priori*.

At present, however, poverty and brutality are increasing worldwide; philosophy – if it is engaged at all – is more and more superficial, and the arts, with few exceptions, are becoming ever more insipid and lowbrow. And finally technology, the progeny of the natural sciences: Technology is totally abused, and everything that can be made is in fact realized. In this manner, people's living conditions are made inexorably worse, a perverse application of technology!

All of this is framed by parents who are no longer up to their tasks, who often do not want to undertake them, and finally the schools, which are only half-heartedly supported. About politics, I will say nothing more here. Everything is made into merchandise, purchasable, and thus registered, even the most complex structures of matter. What has come over us?

In this human development, which must undoubtedly lead to a catastrophe (and recognizing that requires no great acumen), one definitely knows what has gone "wrong," and one asks – or at least the people who are still interested ask – how it is possible, in principle, to escape this disaster.

The explanations presented here hope to make a contribution, although the starting positions lie in the natural sciences, because one thing seems clear: only a removal (or at least a thoroughgoing correction) of the various current worldviews of the religions and their replacement by a cosmic ethics and understanding of the world, which is not based on mere belief but supported by the unvarying insights of natural science and all of its consequences, particularly in physics, offers the prospect, in principle, of leaving humanity's previous path!

To what extent this will be possible, and indeed whether it will be possible at all, is impossible to say. Optimism is not a necessity, but again we would like to realize that the extensions of theory that would be necessary can only come about through extensions of experience, and possibilities for experience, and that these extensions of theory also include the previous insights (theories). This process is how the concept of truth is defined in the framework of scientific discovery of insights, and that is a necessity; we have no other choice, no other possibilities, as has often already been noted!

We will also be able to say that the knowledge gained about the physical world, along with its consequences, goes far beyond what religion and philosophy are capable of offering, and undermines the religions' claim to absoluteness.

That may come as a surprise, and for that reason, wave mechanics should be presented more carefully, in a way that is appropriate to these conclusions. We will thereby recognize that precisely the more far-reaching consequences of our insight into nature, with their cosmic meaning, have to date not been tapped to their full potential. I have the sense that people had not expected consequences of this sort, and that for many people they are unwanted because they contradict and do away with the old, accustomed conceptions. Equation (18) must thus be interpreted further, as it forms the foundation of wave mechanics!

In mathematics, probability is defined as lying between one and zero. $\Delta W = 0$ means that this constellation of particles does not appear in nature, thus that it is not observed. We would achieve maximum probability if we were to sum ΔW according to (18) over all possible spatial configurations of the n + N cubes, as we would then obtain the probability for the location of all particles in the entirety of space. That, however, is always the case, because all particles must be somewhere in space, and so the probability has become a certainty. We may thus say that a probability of one (the maximum value) means certainty.

In mathematics (differential and integral calculus), one allows all of the possible cubes to keep getting smaller, such that their number increases correspondingly, and imagines that all of space is filled with them. This so-called "infinitely" small $\Delta\tau_j$, -> $d\tau_j$, of which there are an "infinite" number, will now be put to use to sum up, in accordance with (18), the "infinitely" small dW. This process is known as integration, and the $d\tau_j$ are called differentials, and one writes it as

$$\int \ldots \int \Psi^* \Psi \, d\tau_1 \, d\tau_2 \ldots d\tau_{n+N} = 1. \tag{20}$$

We do not want to explore this in greater depth, only to show with these brief mathematical considerations that the solutions to the time-dependent Schrödinger equation must also meet the requirements of equation (20), so that the interpretation of Ψ will stand up to scrutiny. If solutions to the equation were found that could not complete (20), then they would not be wave functions, and thus unusable for our purposes.

We also want to mention that the H-operator, which contains, among other things, the interactions of the particles, can also encompass the time-dependent effects of electric and magnetic fields from "outside" the particle system on the particles. H is then dependent on t. Ψ by contrast always depends on time t, and shows oscillation characteristics, which is based on the nature of matter (wave equation).

We will recognize this more clearly later from a simple example; what we see here is actually the de Broglie matter wave again, only taken into the generalized theory. What is decisive is a characteristic of the time-dependent Schrödinger equation, which is that in the search for a conclusion a starting condition for time (t = 0) must be stipulated.

In other words, anyone who wants to solve the wave equation must state what structure the wave function Ψ (or its ΔW) should have for t = 0, what distribution of particles (distribution of probabilities) should be supposed at the beginning. That states the problem: we want to find out about the behavior of the system for t > 0.

The choice of the starting condition becomes a part of Ψ, while H remains unchanged. That means that we are looking for solutions that from the outset display the stipulated initial condition (initial state) for t = 0.

There may be a variety of initial conditions, and precisely that is a certain attraction of the time-dependent Schrödinger equation, in that we may specify what we want to know, which experiences (processes) may be expected, and thus H – along with the conditions on Ψ (t = 0) – provides the appropriate initial condition for the time-dependent Schrödinger equation, from which we draw the desired insights. This is valid for the entire cosmos, as we have seen, because the conclusions of wave mechanics have this characteristic as a part of their nature, as has been shown above.

The attainment of knowledge, like that in the equation in front of us, is the result of much effort and work, and also of considerable understanding and ingenuity because the higher dimensions of space (n, N) lead to considerable mathematical problems. Finding results from the Schrödinger equation requires large and powerful computers, although almost without exception it is necessary to calculate with tilde-Ψ, in which case questions of precision arise. In this area, a large field of research has arisen, in which methods are developed to find solutions, especially approximate solutions.

In 1925, Wolfgang Pauli discovered the principle that was named for him. It places conditions on the wave functions in the form of required symmetries. Without fulfilling the Pauli principle, congruence with experience cannot be established! The requirement arises independently alongside the Schrödinger equation, and cannot be deduced from it! It is valid for particles, as long as these particles cannot be differentiated. In the case of electrons, for examples, Ψ (and also tilde-Ψ) must change their sign if the coordinates (including spin) of two particles in the wave function are transposed.

A very simple example: two electrons on the x-axis

$$\Psi(x_1, s_1, x_2, s_2) = -\Psi(x_2, s_2, x_1, s_1) \text{ (anti-symmetry)} \qquad (21),$$

in which s stands for the previously noted spin coordinate.

For other types of particles, the principle may require symmetry with respect to transposition, which means in this case that when two coordinates are transposed the sign does not change. Pauli discovered the principle through studying the periodic table. It is a fundamental element of wave mechanics!

We see from all of this what is, in the end, required of researchers who approach nature more closely and deeply, who want to understand nature at its heart; that is, who want to develop a theory. It is absolutely necessary that people approach problems impartially; otherwise, the paths toward insight will be blocked or missed completely! That is an important aspect!

What, for example, would actually have happened if humanity had not, despite all the evidence, accepted the uncertainty principle with the argument that all moving particles of matter can have their location and momentum measured precisely and simultaneously, and that had always been true? Even the "ancients" knew that already! If necessary, the followers of the "uncertainty lie" would have been persecuted because they distracted young people from the "facts" and imparted a false "view of the world"!

In this situation, we would never have understood the structure of matter, nor the movements of the particles that comprise it.

The most fundamental experiences with matter would have remained incomprehensible to us, and we would have had to believe that waves and particles exclude one another and yet belong together in a duality. The cosmos would have remained incomprehensible. We would have lived in a world whose reality was not accessible to us, that could not be accessible. "Prophets" would perhaps have attempted to instruct us that the contradiction was a matter of nature, and we should live accordingly! The contradiction would have become a commandment of the cosmos! Maybe in the end people would have repressed one question after another about the meaning of the universe – and of life – and eventually outlawed the natural sciences.

Perhaps also because certain circles would have been able to gain advantages from the contradiction in nature.

I presume that truth is not a matter of indifference to adult and honest people, because their lives are built upon it. In the end, if one is honest with oneself, one wants substantiated knowledge as a foundation for life, as a condition for action, and to be able to share that knowledge with others.

In the case of the dualism of particles and waves, this contradiction is easily resolved with the help of wave mechanics. The time-dependent Schrödinger equation describes the motion of a particle (of a unified type) through the slit and delivers Ψ; from that and behind the slit one receives the ΔW-distribution, which corresponds to the diffraction pattern of the slit, because Ψ is a wave depiction! The values of ΔW then give the probabilities for the presence of the particles behind the slit. In simpler terms: particles remain particles, and their "movements," the probabilities of their presence, are measured by Ψ, also by ΔW, and result from the Schrödinger equation.

Ψ is thus a descriptive magnitude, not measurable, but it determines the diffraction process as a solution to the Schrödinger equation. ΔW, which can be calculated from Ψ, is verifiable. That solves the wave-particle contradiction! For photons, as experiments show, the position of Ψ is taken by the electromagnetic field whose square (intensity) corresponds to ΔW.

As we shall see, the solution of this contradiction has unexpected consequences, and is in harmony with the predictions of the uncertainty principle! (Chapter 11)

We thus recognize that matter shows characteristics that were unexpected, and simultaneously mean the end of the notions of materialism! We will address this more closely, and reflect on it.

It is possible to see already that the discussions about the scientific results, and the consequences that arise from them also touch closely on aspects of religion and philosophy, and must, in the end, incorporate them.

These remarks have actually shown an old experience; namely, that one comes to new insights, even to a "recovery," if one leaves behind comfortable habits and very fixed points of view. It often turns out that these old concepts and comforts were long since inappropriate for our lives, and that our development required a different sort of thinking and acting. In other words: much that is accustomed, habitual and traditional, what seemed so self-evident to us, does not need to be nearly as everlasting as we imagined at the time.

We are speaking about understanding nature, because here, at the direct confrontation between humans and the cosmos, a new aspect arises: unambiguous verifiability is possible, and allows for the elimination of errors! The history of the understanding of nature clearly shows this process. In the end, it was to be expected that this would be the way to be able to express the entire world of matter – and matter is the world, as our experiences have shown – in *one* theory. That keeps errors to an absolute minimum, and enables one to reach the maximal level of truth, in the sense of a comprehensive congruence with experience to date and with the possibilities (probabilities) of expectations.

This situation exists in wave mechanics! There can be no absolute truth, but the insights gained can no longer be changed, only (as demonstrated above) incorporated as such into an extended theory! What has been achieved so far must be taken seriously, and used as the basis for further considerations; the consequences are thus binding!

There is no need for agreements or covenants, nor for revelations; believing things to be true is no longer required, not necessary. The development of humanity can also show additional maturing, leading to new and proven insights; that, however, involves growing responsibility toward society and, in the end, also toward the world. That is often forgotten, consciously suppressed, as if one wanted to rescue a part of earlier adolescence!

We will follow these thoughts, which are directly related to our topic, developing them further and recognizing that the problem of matter is, in the end, the problem of human existence. That is because during our existence we are continuously in close contact with electrons and atomic nuclei, and their effects; indeed we are made up of them. Every process is an expression of matter, and that holds true for the entire universe; under the right circumstances parts of atomic nuclei appear, if particular conditions are met.

Wave mechanics has thus show that matter is not what primitive materialists to this day commonly claim. Many consider consideration of matter to be unimportant, often "beneath the dignity of an educated person"! For everyday life, one often hears, it is not important to know anything about matter – that is aggravating and distressing. We will soon show how mistaken this opinion is, and how much damage it has already done to humanity.

Before we do that, however, we should examine the time-dependent equation more carefully, thereby gaining new insights and fixing some starting points for subsequent considerations.

If the H operator is independent of time t, then the system is not under the influence of a field that changes with time, and the wave function can thus take on this (among other) special form

$$\Psi = \varphi e^{-iEt}, \qquad (22)$$

in which φ does not depend on time; thus (as a starting condition) the following is valid:

$$\Psi(t = 0) = \varphi. \qquad (23).$$

Setting (22) in the time-dependent Schrödinger equation produces

$$H\varphi = E\varphi. \qquad (24)$$

We call (24) the time-independent Schrödinger equation, and it is possible to show that E is the total energy of a system comprising n electrons and N atomic nuclei, which is now independent of time. That means that (22) represents a stationary solution to the time-dependent equation, and in fact ΔW from (22) is also independent of time.

It is interesting to not that, as a general rule, (24) only has solutions φ_k for certain values of E. For that reason, we call (24) an eigenvalue equation for E_k.

$$H\varphi_k = E_k\varphi_k \quad (k = 1, 2, 3 \ldots) \qquad (25).$$

Every value of E_k for the system corresponds to a particular φ_k solution (eigenfunctions of φ_k). Naturally, φ_k must also satisfy the Pauli principle, and $\Psi^*\Psi$ is replaced by $\varphi_k{}^*\varphi_k$, if ΔW is to be calculated. From (22), it follows automatically that the integral over $\varphi_k{}^*\varphi_k$ must also be one, or, as it is known, normalized to one.

Naturally, all of the static states under (24) conform to experience. The states of the nN system have been quantized! Wave mechanics was the first to be able to explain this. However, values of E_k only occur for certain values of nN (including the charge of the nucleus); that is, for example, in accordance with the experience that only certain atoms or molecules exist (see above). Solids and liquids are also described by (24); however, it is then usually necessary to work with the time-dependent Schrödinger equation, which also applies to chemical reactions.

Static states are exceptions in nature; they occur if the system is undisturbed (the starting condition is thus conserved).

At any rate, one can consider as a good approximation two states E_k and $E_{k'}$, in which the system changes from E_k to $E_{k'}$. The difference in energy

$$E_k - E_{k'} = h\nu_{kk'} \qquad (\nu_{kk'} \geq 0) \tag{26}$$

shows, according to the well-known formula, a frequency $\nu_{kk'}$, which is in fact observed at the transition. The $\nu_{kk'}$ frequencies provide the spectrum that is emitted by atoms and molecules, and also the one that can be absorbed (absorption). With the assistance of φ_k and $\varphi_{k'}$ it is also possible to calculate the probabilities of the various transitions (26), although calculations often result in "forbidden" transitions (probability zero), which are then not observed; all of which conforms with experience.

Thus, all of the characteristics and behavior of matter can be described and calculated. In certain circumstances starting values can be determined that allow the characteristics of matter to then be derived. All of this follows from Ψ. All of the calculations of the understood relationships are based on the Schrödinger question, on the Pauli principle, and include the uncertainty principle.

We should add a few explanations and example. If we consider the example of atoms, such as the case in which N = 1 with n electrons and the magnitude of the charge of its nucleus Z, then the Schrödinger equation describes the n electrons in the field of the nucleus with a charge of Z. A neutral atom is represented by Z = n, and it can adopt a particular total energy value E_k (E_k =< 0), which results in the lowest value (ground state of the atom)

$$E_0 \leq E_1 \leq E_2 \ldots E_k \leq E_{k+1} \leq \ldots . \qquad (27)$$

This energy spectrum is characteristic for each atom, and thus using (27) makes it possible to discover what kind of atom (element) is present based on its optic behavior. The theory also offers the potential to influence the environment, if a term in (27) changes. (In this manner we learned, for example, that the sun's surface contained vast amounts of helium.) It is also used, for example, to gain information about the composition of the atmospheres of other heavenly bodies.

In the case where $n = Z = 1$ (and $N = 1$), a hydrogen atom (atomic symbol H) is present. Here there is also an energy scale (as in (27)) with an infinite number of values for E_k that approach one another continuously to the right in (27) until finally E_∞ is "reached." Because the E_k values are the only existing bond energies between the electron and the nucleus (proton), E_0 (lowest energy) is the bond energy of the ground state. At E_∞, we have separated the electron from the atomic nucleus (the atom is ionized) and additional increases of energy can only raise the kinetic energy of the electrons as they fly away. All $E \geq E_\infty$ forms a continuum, which is also captured by the Schrödinger equation.

Similar relations pertain for $n > 1$, although these are considerably more complicated. Here E_∞ is the bonding energy of the complete system of n electrons. If one electron is detached, the bonding energy of n-1 electrons remains, and we have a positive atomic ion. An n+1 system is also often stable, thus forming a negative atomic ion (fluorine, for example, can have both F^+ and F^-).

Values of $N > 1$ start the world of molecules, insofar as the system (molecule) is stable for a given number n of electrons. For molecules, the probability distributions of the N atomic nuclei in ΔW are limited to particular small areas of space that determine the "size" of the molecule, and the electron coordinates are integrated via (18).

With $N = 2$, for example, the distance between the two nuclei (the maximum of the probability distribution) depends on the state E_k, and naturally also on the nuclear charges; this is true for all molecules ($N > 2$). All $E_k > E_0$ are called excited states (as is true for atoms), whereby for each state a different configuration of the atomic nuclei can arise! For each new configuration, there is of course a different probability distribution ΔW. From that, all additional information about matter can be obtained, because the values of Ψ_k are also different.

Consider benzene again: n = 42, N = 12, wobei $Z_1 = Z_2 = Z_3 = Z_4 = Z_5 = Z_6 = 6$, $Z_7 = Z_8 = Z_9 = Z_{10} = Z_{11} = Z_{12} = 1$.

One of the stable states of this system arises from the Schrödinger equation as the flat benzene ring (C_6H_6), in which six carbon (C) atoms form a circle and each carbon atom is bonded to an H atom at the same distance. The system can, however, take on other states, such as three C_2H_2 (acetylene) molecules that can interact with each other. It is also possible to observe six C-H systems and derive them from the equations, down to three C_2 molecules and three H_2 molecules. There are even more possibilities of how the system described above can show itself. Benzene has the lowest energy, measured in terms of the energies of the twelve atoms in their ground states!

What possibilities can be found in protein bonds?

Ultimately, atoms and molecules can interact with each other, which can lead to chemical reactions. To encompass these, we return to the time-dependent Schrödinger equation, in that we embed the molecules and atoms in question into the starting conditions in order to recognize how the reaction process for $t > 0$ takes place. In the process molecules may, for example, fall apart, create new aggregates, or exchange component parts with other molecules.

The rates of the reactions or the speeds of the reactions can also be computed with assistance from wave mechanics data! Gases, liquids and solids can also, in principle, be investigated. Wave mechanics sets a firm theoretical foundation! The same holds true for biological systems, although one should always keep in mind that the values for n and N are not limited!

Only the ability to find a solution poses limits again and again; this, however, is only a technical question, as we have already pointed out. For quantum chemists it remains a hindrance and annoyance to know and constantly experience that we possess a complete theory of matter, but that even with the help of mathematical methods we are reaching the limits of what can be done with computer technology, if we are not able to develop new processes to help us further. That is because as n and N increase, the dimension of the hyperspace in which the particle move also increases, and all of that according also of the Schrödinger equation.

With these brief considerations and examples, we wanted to suggest certain aspects that would demonstrate the current state of the theory of matter; certain details need to be examined more closely.

Strictly speaking, it is mathematical logic that enables us, with the help of an equation (and the additional conditions), to obtain complete information about matter that includes and portrays the entire world, independent of n and N!

In addition to the just-mentioned problem in the field of solvability, there is another conclusion that must be noted, as it is unfortunately often overlooked: a system of logic that is self-contained in its conclusions and free of contradictions – and we are only discussing such a system here – can, with its own methods, neither prove nor contradict these conclusions. These conclusions are known as "undecidable." There can thus be no closed system of rules of thought that confirms itself. This proposition is known from the work of Kurt Gödel, and he proved it less than a decade after the discovery of the uncertainty principle.

I refer to this proposition for two reasons. First, it has a certain relationship to uncertainty, which also does not allow precise values of position and momentum, and leaves this data indefinite. Second, the theory of wave mechanics is free of contradictions and complete in its logic, with regard to matter comprised of electrons and atomic nuclei. Its "proof," however, comes from the requirement for full conformity with experience in the past, present and future. That means, in other words, that the mathematical conclusions (theory) in their totality are "undecidable" in Gödel's sense of the word, but find their certainty through experience and are likewise defined by experience! For precisely that reason, wave mechanics can itself, as it were, confirm what the probability concepts of physics determine, and which can no longer be changed but which can be *extended*, as discussed above. The uncertainty principles are thus recognized as true, and can be incorporated into a generalized theory without losing their validity.

Belief and hypothesis cannot be measured against this fact, especially when they are contradictory but still purport to be concerned with the same problem, and thus claim to give answers to the same questions!

I have taken up this topic on several occasions because it is important to me to point out that we are employing widely differing terms about truth, and the broad majority has, through thoughtlessness, not noticed, but also that this is often not pointed out! The term "probability" also falls under this sloppy usage, and very few people consider its relationship with truth.

Similar superficiality can be seen in the use of the term "matter." For this reason as well, I continue to point out, almost swearing to it, the ubiquity of matter, and also that its inconceivable variety is based on just a few "types of particles, something that should excite, or at least astonish, every thinking person. The opposite is more generally the case, which is understandable to a certain degree, if one considers that matter is the total self-image of our nature, and presents, in a real sense, itself. This intimacy with matter, which includes everything, inhibits amazement about it; everything becomes self-evident existence. That could also be a reason that the findings of wave mechanics were so simply accepted; otherwise, an impartial observer must have had the insight, or at least the thought, that all of the expressions of matter, in any form whatsoever, could be traced back to the workings of electrons and atomic nuclei.

That includes spirit and soul, if one presumes a complete description of "particles of matter," that is, of nature, as will be shown in the next chapter.

In fact, based on all that we know, there is no reason to disavow this opinion, because it can not be evaluated as a hypothesis if we consider wave mechanics critically and exclude arbitrariness. We have no choice! We should keep in mind that matter can no longer be considered materially. It shows much greater freedom!

Chapter 11 – The ultimate consequences

Now we are finally coming to the implications. I would wish (as I have often said), that these explanations would be read impartially and without prejudice, just as, for example, it is expected of every researcher in the course of his or her research – and that is often not easy! That especially holds true for the implications of scientific results! Wave mechanics plays a decisive role; it forces consequences that affect the existence of humanity, and thus affect every individual person!

It is thus both astonishing and alarming that some of these implications, particularly the important ones, have to date not yet been discussed. The implications lend themselves to new ways of life, and special forms of consciousness that would clear away most of the evils and burdens of human life, and thus new possibilities would be shown, ways to discuss and clarify the problems and tasks of a renewed humanity from a completely different point of view.

Let us consider the uncertainty principle (12) again. It reveals that location (x, y, z) and momentum (p_x, p_y, p_z) of a particle with mass m and thus velocity of v_x, v_y, v_z along the x-, y- and z-axes do not both exist at the exact same time in nature. Proceeding from the assumption that the particle with mass m is moving at a speed still considerably less than the speed of light, because otherwise the results are more complex but lead to the same conclusion, then we can write the equations in the following form because (as a result of relativity) m can be considered independent of velocity v

$$\begin{aligned} \Delta v_x \Delta x &\geq \frac{\hbar}{m} \\ \Delta v_y \Delta y &\geq \frac{\hbar}{m}, \\ \Delta v_z \Delta z &\geq \frac{\hbar}{m}, \end{aligned} \tag{28}$$

such that the uncertainties refer to velocity and location (see above). Now this fact, which is expressed in (28), can only be interpreted to mean that the particle has no orbit, because only when a location is precisely known simultaneously with speed and direction of movement can a particle's next location be calculated. That, in turn, means that a particle would follow a path (curve) in space!

However, according to (28), precisely these conditions are not fulfilled, and thus, as noted, no orbital movements of the particle are existent. We thus do not know the particle's next location, if its starting position had been known, and that in turn means that for a particle's "movements," there is no causality, because we cannot predict its next location in space. That corresponds with the conclusions of ΔW, the probability distribution. Thinking further: If we want to find a particle (observation) we must know that the finding (effect), strictly speaking, takes place without a cause, which was already expressed in (28) as the lack of an orbit. Causality, however, requires cause and effect! That means that matter at its innermost scale (electrons and atomic nuclei) is provided without a cause. Wave mechanics has recognized a new reality in respect to matter, and thus for the entire universe, a reality that previously must have escaped our notice because we had not thought so fundamentally about the world and about matter, or if we did it had been in a biased manner.

If, in principle, only effects without causes[4] exist and all material processes are bound to changes in ΔW, which can most readily be recognized in the micro-particles, there can thus also be no first cause, which we had previously assumed to be a cause at the creation of the world! The universe thus has no first cause (first movement), and with that the concept of a creator is no longer tenable – there is no God!

The wave mechanical description and explanation of the cosmos has laid the foundation for a society that is free of belief. The universe can be described as matter and probabilities, and thus be completely understood, because wave mechanics is a complete theory with regard to electrons and atomic nuclei (matter), includes all structures

4 See, for example, interview in *Der Spiegel* with D. Kehlmann (49/2005, pp. 175–176).

of matter, up to and including life in its highest forms of spirit and soul, as we have already pointed out and explained several times. The critical references above concerning the role of the religions are in connection with the insights just obtained, and fundamentally so!

Let us say it again somewhat differently: One may suppose that matter is invested with soul and spirit from the very beginning; there are molecules that presuppose life; and life, in its unfolding, is an expression of the variety of particles in their characteristics and interactions, as they are expressed in the Schrödinger equation, together with the additional conditions, that thus bring about the insight.

Matter has no "material" underpinning; its possibilities are much greater and more varied than earlier philosophers could – or wanted to – imagine!

The lack of belief means dispensing with all of the personal and interpersonal tensions that have shaped and burdened the history of humanity, up to and including wars, exterminations and persecutions, even though many religious reasons and motives are not always easy to recognize! Ideologies must also be discussed in this framework, because they have a great deal to do with belief and make great demands.

This scientifically recognized freedom of belief, as a consequence for people, brings out the equality of persons in a completely new manner that previously seemed unthinkable, because the differences of beliefs had too dominantly influenced the history of humanity and kept it under a spell! We will have more to say on this later.

One apparent objection to these ideas is to note that in daily life – self-evidently – experience causality. This approach, however, does not limit the considerations made above in the least. The uncertainty principles are valid in all cases! We have discussed that above already. The apparent causality is the continuously weakened acausality, and thus effective! We should distinguish among three aspects. For the first, we can consider the relationships in (28) again and recognize, as a further implication, that the right-hand side of the equation becomes ever smaller as mass m increases. That means that the uncertainties have continuously shrinking effects and are increasingly difficult to detect – in regard to causality, however, the expression on the right never goes all the way to zero, so that perfect causality is never reached! Even for larger systems (especially in their innermost reaches) the cosmos remains acausal!

That leads us to the second aspect, which, to a certain extent, supports the first. In many cases, atoms and molecules (with their electrons) decide many processes at a larger scale, in that individual atomic events are amplified and affect macro-scale systems. In these cases, uncertainties in the micro-cosmos lead to uncertainties in the macro-cosmos.

One example is the formation of clouds, which are made up of droplets, which arise from nucleation, which in turn are shaped by the corresponding constellations of electrons and atomic nuclei. The same considerations hold for the formation of crystals. Forms of movement in gases and liquids can, in this fashion, lead to the unexpected formation of turbulence.

Mutations or transformations in chromosomes are processes in the individual molecules in genetic material, and are thus also determined by the uncertainty principles! The exact biological construction of future forms of life – including people – is thus in principle unforeseeable! Predictions may be possible in certain characteristics and aspects, but in many others the processes are simply not calculable, which is also precisely a consequence of the principles of wave mechanics (probabilities).

These are just a few examples; others extend to certain forms of illness and further include the field of decisions.

In general: connections between events past, present and future cannot be fully encompassed by causality. The universe should be seen as a product of coincidence, one that includes all processes with certain probabilities. That all arises because electrons cannot be distinguished from one another, conform to the Pauli principle (in regards to Ψ), and together with atomic nuclei comply with the natural law of uncertainty; that is, behave acausally!

If I hold a stone, a crystal or, for example, seawater in my hand, and consider or perceive how the wind blows across my face, then I have reason for reverence, or at least admiration, but not about the unknown – because most essentials of these processes and structures is not only known but completely captures in a theory – but rather because I live in a world whose inner structure I can, with the help of mathematical formulas, come to know, experience, recognize and understand. I can, in the end, realize that all stones, plants, animals and people are made up of the same "material," which is not so material in its "building blocks," and that makes up the universe, with all of its impulses!

And this universe knows (I believe fortunately) no creator, so that in the end the dangerous old material view of the world cannot be sustained scientifically!

Chapter 12 – The behavior of persons and the physical awareness of nature

The influence of the micro-cosmos is particularly noticeable when larger, complex systems show characteristics that we associate with the term "life." Under pressure from the scientific facts about the concepts of matter, one must presume that this "material" also has the ability to produce consciousness, especially in relation to the impression that the body has about itself – to be conscious of itself. That also includes soul and sprit! In this context, it is particularly worth keeping in mind that the lack of a creator of the world excludes the "insertion" of soul and sprit into a body from "outside"!

This process would otherwise be causal, because it would have been done consciously and with a particular goal in mind, a process with a targeted goal, which is precisely what distinguishes causality! Separation of body, spirit and soul from the insights attained so far is, however, prohibited. (Such a separation would always be tied to a devaluation of the body, because one would be valuing spirit and soul more highly.) The consequences of this separation can readily be followed in history, and we recognize, particularly in the Christian West but also in other religious realms, that the body is denigrated. Fundamentally, everything that is corporeal is "impure" (worthy of disdain), and at the very least is not the basis of "real" life, the more so as the soul should "outlive" the body.

Soul and spirit must, in fact, be functionally defined and recognized as matter; that does not remove their central importance, but they belong entirely to the body and create the person!

Unfortunately, it is a Western tradition that philosophy and religion do not want to give the natural sciences their proper due, which, in the end, is the result of limited prior knowledge. Thus the reputation of mathematics is also poor and deprecated, and insights from natural science are reduced to "incomprehensible formulas." For that reason, to date no great interest has arisen in the insights gained from the natural sciences, and particularly from physics. And that is to say nothing of the implications that result from these insights, which are obviously not clear even to people who are closely involved with physics! Even the pioneers of wave mechanics, at least in my personal impression, do not seem to have engaged with the implications; although of course at the beginning a great deal of work was required to clarify the formal aspects of wave mechanics and to bring them into a proper form.

In fact, however, the implications – as we have already demonstrated – have a deep and human meaning that one, considering it superficially, would never have expected. The exclusion of a "prime mover," a creator – which is a well-founded scientific implication – leads to additional corollaries that determine a clear ethics for all people! That is the topic we will now take up.

The simple fact that the entire cosmos is comprised of electrons and atomic nuclei – which contradicts the old materialistic perception, which insisted on differences – a fact that has been noted here many times and is in the end quite clear to see, must lead to the insight that all of nature is a unity. Spirit and soul grow out of a living organism, life is built upon the structures of matter, which is thus the substrate of everything.

Otherwise, only confusion can be expected, which always turns up in the old conceptions, because these conceptions do not have this broad foundation. Without it, any given conception can be put forward with the same claim to primacy; these confusions lead to tensions!

From that, we first draw one implication, that biological morphogenesis can be traced back to physical (wave-mechanical) ideas and formulations. Together with mutation, self-production and self-organization, as well as the regulation of metabolism, cover the important fundamentals of biological processes. Based on the recognized structure of matter (wave mechanics), intellectual processes can thus also, in principle, be traced back to physical laws.

It is all – and one cannot emphasize this enough – the same "substrate," that comprises a molecule, a crystal, the liquids and gases, the viruses, bacteria, plants and animals, and humans as well. Its characteristics determine the variety of the cosmos. That also holds true

– depending on environmental conditions – for all the intellectual and spiritual achievements and perceptions of the so-called higher forms of life.

The conclusion that results must thus be that the knowledge that we are all "one" in this world, this insight that we only differ in our complexity, should decisively change our consciousness! That should lead to respect, above all for everything that exists in us and around us, and it should result in help and enlightenment for the people who think differently about this principle and who act contrary to it.

If we are all from the same fundamental matter, then we bear responsibility for one another, and for the systems that either depend on us as a result of their lesser complexity or have not yet been able to attain our insights, although clearly for evolutionary reasons self-defense – under certain conditions – can be allowed.

Thus knowledge from the natural sciences – when it is understood and leads to insight – forms the basis of a morality and ethics! Certainly the situation is not quite as simple as it has just been described here, but these considerations and fundamental analyses should essentially outline the task. Additional implications of our scientific understanding of nature will arise from the more precise examinations undertaken below, and these will confirm the lines of thought considered up to this point.

This is where the start of an interpretation the meaning of our existence can be found! This "meaning" can obviously not be given to us from the outside, or reached by coercion; rather, it must arise from within us, insofar as we make an effort to place our environment and ourselves into the context of our scientific insight. Becoming conscious of and engaging with our insights are what form the foundations of our behavior, which leads us to a new form of commonality, one that transcends all of our differences! Looked at this way, the differences are a guarantee of our further development. Based on the commonality, the differences can be recognized, integrated and accepted, without their becoming a reason for divisiveness!

This insight from the natural sciences means that nature is formed out of the one "substrate," which we had once falsely – and thus to our own detriment – called "matter." We all belong together in this cosmos (which we will explicate more closely in the next chapter); only complexity differentiates us, but from that fact – the way "things stand" – no claim to power can be derived! Complexity is thus what increases, from simple molecules to the intellectual realm, and for that reason the systems cannot be pitted against one another. It would always be a fight against ourselves!

The scientifically-based, faith-free society – globally as a humanity without religions – makes possible entirely new and enhances opportunities for ethical living! As the beings with the greatest complexity, we are charged with the protection of all, and thus also with the protection of the development of nature. In humans, matter has become conscious, and observes, so to speak, itself. Humanity is the phalanx of the evolution of matter! That defines us. Morality and ethics are thus binding for all, if we wish to survive in this world.

That is the commandment that follows from the insight that the world is comprised of matter, and is properly described by probabilities. Within this connected framework, which ensures stability, human life must take place. Freedom is thus not some wild abandon, but rather the expression for the experience of the scope for development that the environment, as a result of its structures (and our insights), opens up for us, without harming itself.

To guarantee cooperation within the unity of humanity and nature with the least amount of friction, we must cast aside all of the damaging ballast weighing down our thoughts and blocking the way to insight into the connections of the world. Some of the cathedrals and structures of thought built up in the last centuries or millennia are now on trial in light of the fact that humans are changing the world in a rapid fashion. They are intervening in the arrangement of electrons, atoms and molecules, without having understood this arrangement sufficiently to draw the necessary conclusions! Developments show that no balance between humans and matter has taken place; perceptions of models that were taken seriously worked themselves in between our brains and the matter out of which they are made, while in society accumulating, unjustly comparing, unevenly distributed means of payment (capital) continue to exist. Most people still seek their salvation in fictional, unearthly powers, transferring responsibility to them and expecting thereby a calm for their agitation and an end to their uncertainty. The results are wars, persecutions and massacres! The more that our ability to alter matter and the misuse of this ability increase, the greater the need we have for a harmonious and insightful behavior toward the natural and material contexts of our environment.

The desire for knowledge is a primal part of human nature, but the lust for power and exploitation, mistrust, lies and pomposity lead to the repression of the desire for knowledge, or direct it into one-sided and externally-driven paths. The press and politics play along powerfully here, either out of ignorance or calculation, but the media in general are not entirely innocent. After all, the atmosphere in society – which knows too little and often does not want to know – is so influenced that trivialities almost completely dominate the scene and the meaning of existence gets lost. We have already spoken above about the role of religions.

But the consequences of scientific insights inexorably require that humans – in order to survive – sense themselves, insightfully and judiciously, once more as a part of the earth and of the cosmos, and thus remain its most important, highest and most complicated "system." Humans must be the sustainers of everything, if everything it to go its proper way, and thus to make use of the full possibilities of development that are available within the existing (evolutionary) framework.

In the previous chapters, we have only been able to suggest the problem of the implications, despite a certain breadth and occasional repetitions. More details can be found in the literature that was noted at the start, from which some passages were also taken. I know that most of us – insofar as the topic interests us – reject all of this as unreal, certainly also finding it often fairly naïve. Such a sketch of ethics will be characterized as absurd, but I would like to warn against this error. Because all of this can only be understood is one has properly understood the conclusions of physics.

That is not simple, if it is possible at all, given the general state of knowledge and education (Chapter 13). In a certain sense, the insights must be "internalized;" when these have "taken hold" of us, only then should one think about it and evaluate it. To do that requires knowledge and insight into the scientific connections in a way that can have nothing to do with preservation of power, or with immediate, superficial gains.

Seen from our present situation, such considerations, which illustrate scientific implications –seen superficially – must be viewed as absurd. If we have, however, recognized that this universe has no creator, then we are confronted with a completely new situation. Then the question of the meaning of life will be asked in a way that it never was before, insofar as we are still capable of appreciating the deepness of the question.

In this complex of problems, there is a compulsion to seek solutions and decisions. Because the ramifications are so cosmic, the answer should be too!

The almost simple recognition that a "whole" must be see is not new, but it has not previously been seen in the light of substantiated scientific findings, not that it necessarily defines an ethics. In this process, the current situation of humanity may not influence the decision that is imposed upon it, even if the result completely conflicts with the situation! At the start, the substantiated insight must first be in place (how this insight could perhaps be extended must remain open). Humans are referred to the invariant fundamental questions, and clarity is largely attained. The foundation has been laid; now it is up to us humans. (The corresponding mathematical relations, as they will be discussed in Chapter 13, point toward additional effects.)

When all is said and done, the implications lead to a change of consciousness! The insight forces us to see everything around us differently, and also to classify ourselves differently. Matter is thus a substrate, in which everything that could be and often is realized within the framework of evolution under certain conditions (probabilities), is already trained and prepared. From this perspective, death is the reduction in complexity of the particular individual.

These individuals show what is possible from the material. And that is a further implication of the wave-mechanical results; namely, that every individual, even every being, is a unique formation that can never again be repeated – a recurrence is impossible!

Such a recurrence would in fact mean that in a development completely equal values of nN the same structures between the particles must arise, but that is not yet enough. This system would have to have the same experiences as the "dead" one; otherwise, it would be a different being – which is also a question of the molecular structure. These experiences must include all of the impressions of the senses, all thoughts and feelings. That is, in consideration of precisely the variability of nN systems and their surroundings, completely impossible.

Although electrons and certain subgroups of atomic nuclei (with the same charge, mass and nuclear spin) cannot be differentiated, and all behave acausally, the macro-systems that they form are unique, which is a function of the structure of the nN systems, which can form existing combinations that are impossible to overlook, as described by wave mechanics. Uniqueness thus in "totality"!

Our uniqueness, and the uniqueness of others should compel mutual respect and care because all other behavior can only – in the long run – mean the downfall of precisely the values that have led to our insight into nature! This behavior, this respect, includes the knowledge that death is the complete end of a particular complexity, irrevocable and with no rest, except of course for the remaining matter, whose particles can be included again in other systems! That is the other side of our insight, that everything is one!

In terms of this insight, we should always consider one another, think about one another and feel for one another; and every form of life should be accepted and, where appropriate, included (even if problems sometimes arise, some of which are due to evolution or can be traced to misunderstandings). In any event, we must change our relationship with matter – we have no other choice, because scientific findings are, as has often been noted, unambiguous! In this context and over the long term, a means of payment – especially in the development of its application, as happens constantly – cannot be reconciled with the existence of matter in all of its variety.

With the concepts that have existed previously, which are not well-founded, we have backed ourselves into a corner. Our "progress" is a destruction of material structures on a broad basis and into the highest orders of complexity. And if we believe that we have gained something thereby, it has been considered far too little, and is thus damaging, and could largely be gained by other means, if it is necessary at all! Intellect with decency is far too seldom found! And with that, we close the wide-ranging considerations.

Chapter 13 – Non-locality in wave mechanics

Now is the time to discuss a fundamental characteristic of the nN system that is unexpected, arises from the time-dependent Schrödinger equation, and was not yet addressed in this general form. It ultimately justifies the conception of "everything" and is directly related to the explanations presented in the previous chapter! In that part, the implications were already demonstrated to a great degree, and in particular the unity of the cosmos was emphasized.

These insights cannot be derived from the time-independent equation, because the process that we now wish to consider are themselves time-dependent.

Unfortunately, these results, which can be called cosmic, cannot be presented completely without mathematics, as they are implications of the Schrödinger equation. However, we have already presented the most important and necessary aspect in previous chapters.

First, we break the nN system into two separate systems, A and B, with $n_A N_A$ and $n_B N_B$.

This yields

$$\begin{aligned} n_A + n_B &= n \\ N_A + N_B &= N. \end{aligned} \tag{29}$$

Further, we assume that systems A and B are stable states, and we want to include an additional state (denoted by *) in these considerations. We will designate these systems with the abbreviated notation of A and B, and A* and B*, respectively. Finally, we make the decisive assumption that the two systems are separated by an arbitrary distance.

The question we pose is whether any sort of relationship exists between the two systems. Do A and B "know" about each other, even though they are so far apart that the speed of light plays no role? Does distance play any role at all?

Because we are considering the entire nN system, according to the specifications above, we will proceed by approximation from four wave functions, Φ_j (j = 1, 2, 3, 4), which are comprised of the wave functions for A, A^*, B and B^*. We depict them as follows, in formal notation:

$$\begin{aligned} &\Phi_1: A + B \\ &\Phi_2: A^* + B \\ &\Phi_3: A + B^* \\ &\Phi_4: A^* + B^*. \end{aligned} \tag{30}$$

That means, for example, that Φ_2 is comprised of the wave functions for states A* and B. We further assume that the wave functions for the four states are exactly known. It is possible to show that Φ_j, which describe the complete nN system are thus also exactly known! With these four Φ_j, we shall then approach the solution of the time-dependent Schrödinger equation by approximation; that is, obtain Ψ-tilde. Incidentally, the energies of the AB systems are also of course known, which we designate $E^*{}_A$, E_A, $E^*{}_B$ and E_B.

It is necessary to emphasize again that the sizes of the AB systems can be chosen completely freely and are incorporated in the operator H. A and B can be molecules, crystals, or even protein bonds, or parts of the material that are made up of electrons and atomic nuclei! (Complexity plays no part in this consideration.)

That the Φ_j are known exactly – including the complete energies of the nN systems that belong to them – is an assumption made to be able to complete the calculations, but it is also justified in physics because these solutions exist in reality; they are – because they describe stationary states – solutions of the time-independent Schrödinger equation. Our aim, however, in accordance with the question that we have posed, is to capture time-dependent processes between A, A^* and B, B^*!

Because the two systems are stationary and separated by a very great distance, they can be depicted as follows:

$$\begin{aligned} &H\Phi_1: (E_A + E_B)\,\Phi_1 \\ &H\Phi_2: (E_A^* + E_B)\,\Phi_2 \\ &H\Phi_3: (E_A + E_B^*)\,\Phi_3 \\ &H\Phi_4: (E_A^* + E_B^*)\,\Phi_4. \end{aligned} \tag{31}$$

In this form, H is the Hamilton operator for the complete nN system. Φ_j complies with the Pauli principle!

Following the standard procedure, we substitute for the approximate solution $\tilde{\Psi}$

$$\Psi \approx \tilde{\Psi} = \varphi_1 e^{-iE_1 t} + \varphi_2 e^{-iE_2 t}, \tag{32}$$

in atomic units, although initially φ_1, φ_2, and E_1, E_2 are not defined, and if we require that Ψ-tilde fulfill the Schrödinger equation as closely as possible (the least deviation from Ψ), we derive the conditions

$$S_{12} = H_{12} = 0 \qquad E_1 = \frac{H_{11}}{S_{11}}; \qquad E_2 = \frac{H_{22}}{S_{22}}. \tag{33}$$

From (33), φ_1, φ_2, E_1, E_2 can be calculated.

Individually,

$$S_{12} = \int \varphi_1 \varphi_2 d\tau\,; H_{12} = \int \varphi_1 H \varphi_2 d\tau\,; H_{11} = \int \varphi_1 H \varphi_1 d\tau\,; H_{22} = \int \varphi_2 H \varphi_2 d\tau, \tag{33a}$$

if it is integrated across all particle coordinates. *(Equations (33a) through (42) are not necessarily required for what follows; they are instead intended to give the mathematically interested reader further explanation.)*

To meet the conditions of (33), that is, to calculate φ_1, φ_2, we insert the functions Φ_j, and write in their place

$$\begin{aligned} \varphi_1 &= \alpha_1\Phi_1 + \beta_1\Phi_2 + \gamma_1\Phi_3 + \delta_1\Phi_4 \\ \varphi_2 &= \alpha_2\Phi_1 + \beta_2\Phi_2 + \gamma_2\Phi_3 + \delta_2\Phi_4 \end{aligned} \tag{34}$$

That is the decisive step, because it is now a matter of calculating the coefficients in (34) in place of φ_1, φ_2, because Φj are known! Setting (34) into (33), one obtains the equations

$$\begin{aligned} S_{12} &= 0 = \alpha_1\alpha_2 + \beta_1\beta_2 + \gamma_1\gamma_2 + \delta_1\delta_2 \\ H_{12} &= 0 = \alpha_1\alpha_2\left(E_A + E_B\right) + \beta_1\beta_2\left(E_A^* + E_B\right) + \gamma_1\gamma_2\left(E_A + E_B^*\right) + \delta_1\delta_2\left(E_A^* + E_B^*\right), \end{aligned} \tag{35}$$

(because the Φ_j are so-called orthogonal functions) from which one may then – if the energies are known – calculate the coefficients, according to which E_1 and E_2 are determined by (33)! These so-called nonlinear equations can be solved relatively simply, particularly because additional conditions are imposed on the coefficients. From $\tilde{\Psi}$ under (32) follow after further calculations (in atomic units)

$$\Psi^*\Psi = (\varphi_1 + \varphi_2)^2 - 4\varphi_1\varphi_2 \sin^2\left[\left(\frac{E_1 - E_2}{2}\right)t\right], \qquad (36)$$

with φ_1, φ_2 according to (34). With (36), ΔW can be easily determined. Equation (36) is the one that we have been seeking, if we additionally write in φ_1, φ_2 according to (34), and if we know the values of the coefficients! E_1 and E_2 must also be known, which is in fact the case (see above), if the coefficients are also known.

First, we recognize that the initial condition (t=0) produces

$$\tilde{\Psi}^*\tilde{\Psi}|_{t=0} = (r\Phi_1 + s\Phi_2 + u\Phi_3 + v\Phi_4)^2 \qquad (37)$$

if we abbreviate:

$$r = \alpha_1 + \alpha_2 \;,\; s = \beta_1 + \beta_2 \;,\; u = \gamma_1 + \gamma_2 \;,\; v = \delta_1 + \delta_2 . \qquad (37a).$$

Now we must select the values in (37a), because we must set the initial conditions! In that choice, however, we do not have a great deal of freedom if we want to study the relationship between A and B. We could, for example, at t=0 a B system in the condition E_B, and in the A system, allow E_A and E_A^*. That means

$$u = 0 \;;\; \gamma_2 = -\gamma_1 \qquad v = 0 \;,\; \delta_2 = -\delta_1 \qquad (38)$$

and (37) becomes

$$\tilde{\Psi}^*\tilde{\Psi}|_{t=0} = (r\Phi_1 + s\Phi_2)^2 \qquad (39)$$

Now we must also require that $\tilde{\Psi}$ is set to one, as stipulated in (20), and that leads to the condition

$$r^2 + s^2 = 1. \qquad (40)$$

With that, the free coefficients in (35) are very constrained. More precise calculations show that, out of everything in the pre-condition equation (33), only α_1 and β_1 need to be calculated, while γ, δ, r and s can be freely chosen, whereby γ_1 and δ_1 are, however, constrained in the values that they may have if α_1 and β_1 are to have real values.

The values of r and s are likewise constrained by (40). The four values for E are also incorporated into the $\alpha_1\beta_1$ conditions, and thereby determine $\alpha_1\beta_1$ and $\gamma_1\delta_1$. $\alpha_2\beta_2\gamma_2\delta_2$ result from (37a). From (36), we recognize that for t > 0 (following the initial condition) a fluctuation in the probability ΔW arises, if $E_1 =/ E_2$! That is a standing probability wave that arises throughout the space, and whose amplitude is given by $4\varphi_1\varphi_2$, which is a function of all the coordinates of the particles involved. More precisely, one must say that the amplitude depends on Φ_j. This standing wave arises almost instantly after t = 0, because the computation of E_1 and E_2 generally produces fluctuation frequencies that are on the order of millions of GHz. Following each period, the initial condition arises again.

The standing probability wave does not appear if $E_1 = E_2$. The computations show that this is the case for the values

$$\begin{array}{lll} & s = 0 & r = 1 \\ \text{oder} & s = 1 & r = 0 \end{array} \tag{41}$$

which correspond to a stationary state for A. Thus, r and s are the mixing parameters for the two states of A. With this step, we have obtained the decisive implications of the time-dependent Schrödinger equation!

In fact, the notion that the universe is one is now physically demonstrated, because there are influences between all parts of matter, which have an effect across arbitrarily large distances, and which can take effect almost instantaneously if the corresponding preconditions (at least in one system) are present (e.g., entangled states). The influences mediated by the probabilities are triggered by the emergence of fluctuating probabilities throughout the space (probability influence, PI). These are, as noted above, standing probability waves, which must be kept in mind.

This affects all systems that are present in the Schrödinger equation (we have only considered two here, in order to show what is most important). It could, however, be an arbitrarily large number of systems of an arbitrary size! The element PI is always present! That is the characteristic of "matter" in the universe! Variable initial conditions lead to similar results!

Anyone to whom that does not give pause for thought, particularly upon considering acausality, which acknowledges no creator, is probably beyond help, and unlikely to realize that the world is different from what people had previously thought and believed to be true! It is even possible to consider the amplitude $4\varphi_1\varphi_2$ more carefully, and according to (36) produce

$$-4\varphi_1\varphi_2 = 4\left\{\varphi_1\left[\varphi_1 - \left(r \cdot \Phi_1 + s \cdot \Phi_2\right)\right]\right\}, \tag{42}$$

and thus realize that all combinations between states of A and B appear. The probabilities are determined by $\alpha_1\beta_1\gamma_1$ and δ_1. The fundamental process is shown as an expression of the non-locality of wave mechanics!

That means perforce that the universe is non-local. In this way, our understanding of nature has attained a deeper meaning, and in this context we refer to the discussions in the previous section. Furthermore, we recognize that because of non-locality certain excited states can be "transported" between the systems, as can be seen from (42).

It must also be emphasized that these results have already been partially experimentally confirmed, although the research (theory) is still at an early stage. The fundamental principles have, however, been identified. Between the material parts of the cosmos, as subsystems, there exist probability influences whose further possibilities must be studied more deeply! Such insights are more than what was previously available to humanity, when people still addressed and engaged fundamental questions with belief. In this manner, a level of truth has been attained that was first possible with wave mechanics!

We should also not forget that in the cosmos – according to everything that we now know with certainty – an uncountable number of PI effects transpire, which we hardly notice consciously, although we know that these effects always, in some fashion, can lead to changes (probability), which then vanish again but which can also remain, sometimes for longer periods of time. If we then further consider that the universe cannot be a conscious creation, one should by no means see this recognition as a devaluation of the world; quite the contrary: all possibilities are contained in a coincidence, perhaps even more than would have been the case if the precondition had been created by a decision (or an objection). Our insight into the cosmos, which we have achieved for ourselves with the help of the natural sciences, opens for us a new freedom, which did not previously exist. At the same, it time lays out morals and ethics that are based on the cosmos itself, and not merely psychically localized by a belief!

Thus we come to a decisive situation, because it is now time to fill the freedom that was achieved on the basis of science, to arrest and ultimately suspend a development that we gotten ourselves into in recent centuries. That things can no longer continue as they are must have long since been clear to people with insight!

We must now presume – as we have discussed above – that we cannot, over the long term, regard "matter" as a means of payment and with a mentality of possession; we ourselves are a part of this substrate, and would thus, over time, also become exchangeable goods and registered property! That contradicts the ethics of the cosmos and must consequently, as one can clearly see today, leads to destruction. The proper behavior, the solution, is at hand: Matter itself has, together with evolution, found one and continues it again and again. That the implications have to date seldom been recognized is mostly due to the low level of human insight into, and recognition of, nature.

That is because not all flows of information between very complex systems on earth necessarily stick to the truth; information has also become a tradable good and serves the increase of profit and property. This greatly endangers the development of humanity, and furthermore is held in the hands of a small number of people.

The larger and more complex material structures become, the more energy is required to keep the conditions and processes intact, and this energy can only be taken from the energy radiated by the sun! That is the way that humanity-free nature works!

In the human species, by contrast, something new arises. Humans leave the ancient path of energy production and preservation of the species. They plunder the planet's reserves, aided in the process by the existence of methods of payment, and an increasingly stronger and more effective technology, which drives everything immeasurably, particularly given the tendency toward unusually strong reproduction. The appearance of new ways of behaving toward nature is, as such, not unknown with the rise of new species, and often led to the stabilization of material development, even though the framework of evolution made no guarantee of existence.

With humans that is obviously leading to their destruction! Rationally, human behavior should be measured by the ethics and morals that have been recognized by science in the cosmos, according to which human abilities have reached a certain height and breadth. These ethics and morals stand in total contradiction to current behavior! The commonality with the cosmos has been lost. Unquiet greed can be seen everywhere; everything has been and will be sacrificed to the market! Attempts are made to make everything for sale; mistrust is the most widespread feeling, because the restraints have been cut loose; everything that can be thought of must be made and brought to the "market." People want to enjoy, at any cost. Repression, domination and exploitation of humans and nature are a source of joy. Hypocrisy and lies are the necessary consequences. Vanity dominates the rules of life, and all of this spreads itself increasingly intensively across the whole planet.

The results of scientific research in the context of the structure of matter are recognized by few people and are not taken into consideration. They are in fact often looked down on, because increasing population has made ethics and morality largely content-free.

For precisely this reason, I have presented the implications of scientific insight in such detail and with so much repetition. The world is matter and probability, and as banal as that sounds, its implications have a depth and breadth for all of humanity that has never been reached before! Not accepting them leads to separation from the river of life, out of the development of complexity, and leads first to inhumanity, with the consequences described above. The high complexity, which has appeared for the first time in humans, requires the broadest possible analysis of these facts!

But how can humanity get free of this vicious circle? Although this question is not directly a part of the topics under discussion here, we should address it at least in principle, because at the end of the day the implications of scientific results are part of its context. It is self-evident that the solution will be found in the scientific/technical field, and if the conditions of life must be fundamentally altered, one can be reasonably certain that an alteration in consciousness will also take place. Tremendous common efforts will be necessary because the solution must be planetary, and affects all humans, but that could also be what causes it to fail, because the current path of development is so far advanced!

I am convinced, that in all cases PI effects also play a role, because what we already know about it speaks in favor of this phenomenon exercising important influence, particularly in very complex systems. I tend to see PI as a "morphological field" with probability character. That is because changes in the electron systems almost always mean that there are also changes in the spatial probability density of the nuclei; that is, a change in the "nucleus geometry," and thus the "forms" and the internal composition of the systems at hand.

Whoever consciously considers this area of our experience can hardly escape this conclusion. The few experiments that already exist between humans and devices confirm an influence, and it is desirable that wide-ranging research should be conducted in this area.[5] As we have seen above, it is a question of an effect that is guided by probability, which must always be noted because it is so fundamental.

For the solution of humanity's current problems, I contend that a networked, fully automated trade in goods and services that is propagated throughout the world and developed in a multifaceted way, powered entirely by solar energy (including stored solar) – although global population growth poses a particular challenge – will in the end – after a long process – abolish money-driven society and thus end the age of capitalism!

This fully automated trade in goods should also be required to be self-stabilizing, similar to the way that a biological network is constructed and correspondingly in contact with the environment. Before this network, which will, among other things, take over all the functions of agriculture and industry, all persons are equal in the sense of cosmic ethics! That means that such a world-spanning network

5 Sheldrake, Rupert. The Presence of the Past. New York: Times Books 1988

stands ready to serve all people in the same fashion for their self-control and self-preservation, and uses only solar energy.[6]

The system thus includes all means of production, which will probably be simplified and generalized, because there will be no competition! All work that had previously been paid should increasingly be taken over by this system. From the previous forms of work, now unpaid, there remain voluntary and desired activities – callings; activities, that is, that belong to individual human nature and show numerous variants, according to the type of person with general and specific capabilities or values – because people want to be gainfully employed, but not just "working"!

In this context, I am thinking about medicine and teaching, among others, about law or research, but also about some other activities that will continue to prove themselves indispensable, e.g, in the context of chronic illness and care for the elderly. Certainly, however, lifestyles will change fundamentally, so that today we cannot say with any certainty what will remain and what will appear in a new form. What is important is that the offerings of the fully automated system are available equally and free for every person!

I am convinced that by contemporary standards – and after a longer period of time – general modesty will prevail, because the envy, showing off and pomposity will no longer have an effect in society; egoism will also change and retreat to its natural form. Incidentally, the transition period is critical, and its development can call everything into question. It is necessary to discuss this much more, if people are determined to address the matter at all – at any rate, much time and many ideas are needed, suggestions and critical advice from all sides!

It could be humanity's greatest task, and some states that will later be dissolved in the commonwealth should take the lead. Even partial solutions are conceivable at the beginning. At any rate, what is presented here is only meant to demonstrate the general outlines.

---

6 Alt, Franz. Die Sonne schickt uns keine Rechnung. Munich: Piper, 2004.

What is certain, however, is that a completely new type of person will emerge, because many terms, ideas and feelings will lose their meaning. Different qualities will arise, that are part of human capabilities, so that by these means quite naturally morals and ethics will arise that give more room for scientific results.

Of course I know that such thoughts must seem completely utopian and absurd – maybe even farcical – to a capitalist society! But it is not a paradise that is under consideration here, and this description should only sketch out a few thoughts and concepts, because some parts can perhaps be changed, added or discarded. What counts is the direction!

The basics should be depicted along with the implications of the scientific worldview. I am, however, fully convinced that the application of extremely technical full automation with the complete employment of solar energy is basically the only way to avoid the oncoming catastrophe.

The insight that the world at its "deepest" is acausal and that this characteristic also has effects in the macro-cosmos, and that the elements of matter can influence one another – in a way whose details we do not yet grasp but whose important structures we have understood – should be reason enough to re-think everything, particularly because to date we have been proceeding from false premises. We are obviously moving in the wrong direction, as history, with its current effects, shows.

Even if these considerations seem utopian to most people at the outset, it is necessary to speak about them, to bring out new arguments, and to have a clearer view of the future. Engagement with them will make us more certain and more mature, which should indeed be the purpose of research, but which is so little recognized at present. That is because today the results of scientific insight serve almost exclusively to build up centers of power, to strengthen the inhumanity and uncertainty that overshadow and limit – or very often even reverse – the positive side of natural science.

Chapter 14 – Concluding thoughts

As we have already discovered above, the origin of the universe should be seen as an act of coincidence, which we can derive from the fact of the non-causality of the elementary particles of matter. Accordingly, there can be no "meaning of creation" – the universe has no purpose! However, we humans should on no account interpret this fact as a catastrophe. On the contrary, this realization leads to freedom in regard to insight into the cosmos.

It is possible to take various positions toward this insight, particularly with reference to our code of behavior. Seen historically, long ago we used the belief in certain concepts to forge commandments of the most varying sorts, according to which we should live, under the threat of punishment that would come sooner or later, perhaps even after death. These various forms of believing things to be true tolerated neither opposition nor discussion about their origins, because belief acts as if it absolute. Nevertheless, critical discussions have never stopped.

The difficulties, in reference to opposition, arose rather quickly from the divergent justifications of these worldviews, so that numerous variations of these conceptions developed, which tried, and continue to try (with cults, among others), to control people's lives. That led to struggle and war, terrible and fanatical! That is, so to speak, any and every religion's "declaration of bankruptcy."

On the other hand, it must be recognized that people – as the most complex systems – view the question of what kind of world they find themselves in as a challenge, or at least (in the beginning of their development) as a stimulus that they kept reacting to. That may well have been the original cause for the development of philosophy, and later of science, disregarding fallacious "variants" up to and including sorcery and superstition, which we only just want to mention. A different path, as we have said, lies in using well-founded scientific insights to bring order to our experiences, and from that to discuss implications for our behavior. This yields the further thought that a random cosmos can in fact only exist if the non-causal particles are of a kind whose interactions can secure stability, at least for a certain period of time.

The characteristics of electrons and atomic nuclei have been reviewed and discussed above. We must presume that these laws of interaction are also subject to randomness, because we can also imagine other interactions (which, however, have not come to pass) that would also guarantee stability. In other words: even the so-called laws of nature are subject to contingency in their origin, and are thus, finally, non-causal.

We must certainly presume that these "laws of nature" could have shaped the possibilities available to complexities that later arose, but it is idle to pursue this metaphysics; through research, we have recognized the laws of nature, the contexts of our world, and we must live with them, like all nN systems that have arisen from them and which will (hopefully) also arise in the future.

It is thus irrelevant – although it is often claimed – that people have an inborn tendency toward metaphysics, ultimately to the divine, which only seems justified if one has heard nothing of the scientific method or was completely uninformed about it, particularly in regard to the results of contemporary scientific research. Sometimes one even has the impression that certain circles consciously ignore or falsify the results; these days that must be called a dangerous behavior. We now recognize in the laws (PI) found in the cosmos, that we must derive from them a code of behavior, a basis for existence, for everyone if we are to preserve our lives and the existence of our species. Based on our scientific insight, we are thus placed in a position to hinder or dissipate emerging undesirable developments. The goal, however, is clear! A "fully automatic economy based on solar energy" must be the foundation of our lives so that the freedom thus won can be used advantageously and for the further development of humanity. Those are – as we have said many times already – the implications of the fact that the world is comprised of matter and probability, as we now definitively know, given that the theory has encompassed all possible experience.

In closing, it is necessary to allude to an apparently trivial issue:

The time-dependent Schrödinger equation can naturally also be written in the form

$$0 = 0 \tag{43a}$$

$$H\Psi - i\frac{\partial\Psi}{\partial t} = 0. \tag{43b}$$

That is astonishing, because at first one would consider the first depiction completely self-evident and insignificant. It is possible, however, to see (43b) very differently: apparently the "null" on the left side of equation (43a) is depicted "split into two," so as to derive (43b), and what is then on the left side – this division – describes all nN systems! Equation (43b) is this a very special depiction of the null, the "nothing," as one might say. In other words, something that is informationless only begins to send signals when it is rewritten with non-disappearing quantities, in which this rewriting (splitting into two) must take place in such a way that the division correctly describes experience, as has often been discussed above.

Now it can be countered that a theoretical encompassing of experience is always done with formulas, i.e., equations, because otherwise no information can be gained; that is, however, in harmony with the considerations above.

How does that now fit with the uncertainty principles, which are inequalities? According to (13), we wrote above

$$\begin{aligned} &\Delta x \Delta p_x - \hbar \geq 0 \\ &\Delta y \Delta p_y - \hbar \geq 0 \\ &\Delta z \Delta p_z - \hbar \geq 0, \end{aligned} \tag{13a}$$

which also includes the equality sign.

There is, however, a difference between the time-dependent Schrödinger equation and the uncertainty principles. While on the left side of (43b) we were able to preserve the wave function Ψ, which fully describes the behavior of nN systems, the inequalities in (13a) state a further proposition as to how the probability statement from ΔW according to (18) behaves in individual cases (qualitative proposition), and that can definitely take the form of inequalities. Equation (36) an also be viewed in this manner. It is thus possible to say that the uncertainty principles make the information about Ψ concrete in regard to location and momentum.

The extended considerations concerning the Schrödinger equation, as suggested above, can thus also be seen as an implication of wave mechanics, especially as the do not contradict the theory. I tend, however, to the supposition that considerations of the "null" offer a fundamental interpretation of the theory, are general, and thus must be understood as a statement of truth in the sense of our insight into nature.

I will say again that only *one* theory can be formulates as the substantiated insight into nature, if one begins mathematically from a decomposition of the "null," which is based on experience. The "nothing" becomes the "divided nothing," and the equations of theory that result from this describe the facts! For the present, nothing more can be said about this except that it makes a contribution to our insight. I hope that astrophysics in particular may be able to confirm this insight; some evidence suggests that it may.

(We should not count ourselves among the "bugs" who crawl over the 3 in the equation

$$-1 + 3 - 2 = 0 \tag{44}$$

and claim that there is something overall – in their cosmos (44) – there which cannot be compensated.)

Thus at the end it remains to state again that the necessary implications of scientific insight must lead to a society free of belief, and thus also necessarily to a morality and ethics that arise from the certainty that the universe is a unity, comprised of matter, that has a degree of uncertainty as a matter of natural law, and which can thus only be encompassed in probabilities. In addition, there is the insight that there are influences (PI) between all parts of matter in the universe that occur almost instantly and independent of the distance between the parts; these, too, are shaped by probability.

The meaning of PI is difficult to recognize because of its structure, but it can nonetheless be exhaustively described! At any rate, no part of matter can be observed completely "alone" and isolated, nor can it feel that way, although gradual differences may come to exist.

PI is active continuously and everywhere, but we cannot always recognize and experience it, because it is a matter of probabilities.

It reminds one of the collective unconscious and synchronicity, as they are described and discussed in the psychoanalytic writings of C. G. Jung. PI could, however, be necessary as a basis for the understanding of telepathy.

On this basis, a change in the contemporary structure of society is absolutely to be expected, because the freedom of belief[7] must be accompanied by a moneyless society, if everything is to fit together and thus establish planet-wide stability. That is fundamentally possible – and is becoming more pressing – if some form of fully automatic society of goods and services is sought after, which can only be based on the fully automatized, and completely global, use of solar energy.

This energy is available to all people, and a development of this sort must arise from society; otherwise, the transition is difficult to conceive, and precisely that is the most critical phase, because the change in consciousness must begin here.

I have wanted to just sketch some aspects of such a society, without being tying it to a particular path, because many ways could lead there. Even the desired goal is only indicated in its most important outlines.

Nevertheless, the indispensable path toward humanity's future is clear: only the complete application of solar energy can supplant the period of late capitalism. The methods of acquiring energy are exploiting the planet self-destructively – often with great difficulty but also with great financial success – and all the time there is an enormous fusion reactor in the sky! This reactor is *giving* us its energy, and how

7 Onfray, Michel. Wir brauchen keinen Gott. Munich: Piper Verlag, 2006. ((Translator's note: This appears to be a translation into German of *Traité d'athéologie. Physique de la métaphysique.* It is up to the editors whether to use the original French, adding bibiliographical information, or researching whether there is an English edition.))
"Glücklicher ohne Gott," Der Spiegel 43 (2006), pp. 188-90.
Dawkins, Richard. The God Delusion. New York: Bantam, 2006.
(I consider the contents of the quoted books and articles as insufficiently grounded.)

we use it is just a question of our technology, that is, of the results of our natural sciences, because the supply is superabundant, while its management is understandable and just a question of our intelligence!

The creation of solar energy in a completely property-less space, and its unvarying and largely equal distribution across the total surface of the earth in more than sufficient quantities creates, in principle, a unique opportunity to construct and automatize self-sustaining systems for the production and distribution of goods without the interposition of a method of payment! This is undoubtedly the greatest challenge to humanity!

With these obviously insufficient remarks, I will leave well enough alone; I am not at all sure whether the global transition can be found and implemented, and then established "forever." Resistance thrives on the desire for power, from habit, and certainly also from the inability to understand the necessity of the new. Adding financial aspects will only further prove the point! Moneylessness? Unthinkable!

Are, perhaps, unconscious desires for a collapse playing a role here? What is certain, by contrast, is that the self-made collapse of humanity is hard-wired if we do not succeed in changing the entire energy economy into cost-free use of solar energy! That is actually easy to foresee. One only needs to extrapolate current developments.

The change would be destined to fail if the sun remains an object of commerce and the network is split up!

The end will be terrible and degrading! I am certain that in those circumstances, people will still call on their gods!

The cruelty of the end will be even greater because differing parts of the world will be affected differently, but everything will be headed in the same direction to the same "final results." Whatever would remain, will be comfortless and hopeless, because the consequences will take effect very late in some areas! The non-local world is in fact "acausal matter" and must be understood through probabilities – and not in any other way. *To recognize that is our call and our salvation.* Why else do we have natural science?

Es gab niemals Gott!

Heinzwerner Preuß

Zwingende Konsequenzen aus gesicherten Ergebnissen der naturwissenschaftlichen (physikalischen) Forschung, insbesondere der Wellenmechanik. Das schließt eine Ethik ein, die darauf aufbaut, s. Literatur[4]).

Zuerst ist die Klärung und die Definition einiger Vorstellungen und Begriffe notwendig, mit denen man in den Naturwissenschaften arbeitet, um Mißverständnisse zu vermeiden. Die physikalische Forschung als Grundlage der Naturwissenschaften basiert auf Erfahrung, also auf Meßergebnissen an definierten Größen, die im Kosmos vorliegen. Diese Meßpunkte stellen in ihrer Gesamtheit das Wissen der Physik dar. Das Ziel der Forschung (Grundlagenforschung) ist allerdings die Verknüpfung dieser „Meßpunkte" untereinander im Einklang mit der Erfahrung. Diese Verknüpfungen werden als Einsicht oder Erkenntnis bezeichnet.

Je breiter und allgemeiner diese Verknüpfung gelingt, desto größer ist der Wahrheitsgehalt der Erkenntnisse dieser Verknüpfungen anzusetzen; diese können nur mathematisch formuliert werden und stellen die Theorie der Physik dar. Damit steht Theorie als Vernetzung der Erfahrung unmittelbar mit dem Wahrheitsgehalt unserer Erkenntnisse in Verbindung, repräsentiert sie sozusagen.

Die Vernetzung der Erfahrungen stellt unsere Art der Naturerkenntnis dar, wobei die Meßpunkte die Grundlage bilden!

Durch die mathematische Vernetzung werden somit alle möglichen Meßdaten erfaßt – damit ist in der Erfahrung Vollständigkeit erreicht. Es charakterisiert die Theorie, daß in den mathematischen Beziehungen keine numerischen Werte der Meßgrößen auftreten, sondern sich durch Rechnungen ergeben.

Grundlagenforschung ist also Erweiterung der Erfahrung, und damit verbindet sich eine Vergrößerung des Vernetzungsbereiches – der Mathematisierung! Die Erweiterung verläuft daher in der Form, daß der alte erfaßte Bereich sich als Spezialfall des Neuen ergibt. Das definiert den Wahrheitsgehalt der gewonnenen Erkenntnisse!

Gelegentlich wird versucht, sichere Teilvernetzungen auf neue noch nicht erfaßte Bereiche zu übertragen, was man Modellbildung nennt. Wie zu erwarten, tragen diese Versuche nicht zum endgültigen Verständnis des neuen Bereichs bei, sondern irritieren oft. Modelle liefern keine Einsicht und sind nicht die Realität der Erfahrung, aber sie können Vermutungen aufkommen lassen, wobei wir dabei die Hypothese erwähnen müssen, die jedenfalls mehr als Spekulation ist.

Als Materie müssen wir die Gesamtheit aller Elementarteilchen im Kosmos ansehen, die jeweils durch ihre Masse (m) und gegebenenfalls durch ihre Ladung (L) oder magnetischen Eigenschaften (M) charakterisiert werden. Allerdings treten in der Natur nur einige bestimmte (m, L, M)-Werte auf. M erfaßt den Eigendrehimpuls und das magnetische Moment des Teilchens. Drei Teilchentypen spielen eine besondere Rolle, da sie in der Lage sind, höhere komplexe Strukturen von Teilchensystemen zu bilden! Es handelt sich dabei um das Elektron mit der Masse 0,000 000 000 000 000 000 000 000 001 Gramm und der negativen Ladung (Elementarladung) e: -e = 0,000 000 000 000 000 000 1 Coulomb. Das Elektron besitzt noch einen bestimmten Drehimpuls (Spin). Alle Elektronen sind nichtunterscheidbar!

Die anderen beiden Teilchen sind das Proton und das Neutron von jeweils 1836-facher Elektronenmasse. Das Proton trägt eine positive Elementarladung. Beide Teilchen bauen die Atomkerne auf, wobei die Anzahl der Protonen die Ladung des Atomkerns bestimmt. Die Masse des Atomkerns wird von Protonen und Neutronen gestellt.

Ein Atom besteht aus einem Atomkern, um den sich eine bestimmte Anzahl von Elektronen bewegt. Ein Atom ist neutral, wenn die Anzahl der Elektronen den Protonen im Kern entspricht. Die Kernladungszahl Z (Z = Anzahl der Protonen) bestimmt das chemische Element. Durch die Wechselwirkung der Atome ist die Bildung von Molekülen vorgezeichnet, somit sind alle Möglichkeiten der Chemie und Biologie gegeben, die daher zu den Strukturen der Materie gehören.

Diese Entwicklung schließt auch lebende Systeme mit ihren eigenen inneren Wechselwirkungen ein, wie wir noch erkennen werden. Die Komplexität der Materiestrukturen spielt somit eine entscheidende Rolle bei der Diskussion des Lebens überhaupt.

Die Vernetzung der Erfahrungen mit den genannten Elektronen und Atomkernen hat zu tiefen Einsichten geführt. Die vorliegende Theoretisierung (Mathematisierung) ergab die zeitabhängige Schrödingergleichung, aus der sich prinzipiell alles berechnen läßt, soweit es die derzeitigen technischen Möglichkeiten zulassen. Das Verhalten der Materieteilchen wird besonders durch eine mathematische Beziehung beschrieben, die als Heisenberg'sche Unschärferelation (1925) bekannt ist. Danach sind Ort und Geschwindigkeit (Impulse) eines Teilchens nie gleichzeitig exakt bekannt (meßbar).

Bezeichnen wir die drei kartesischen Ortskoordinaten des Teilchens mit x, y, z und die entsprechenden Geschwindigkeiten mit v_x, v_y und v_z, so lautet die Unschärferelation nach Umordnung bei fixiertem m,

$$\Delta x \, \Delta v_x \geq h/m$$
$$\Delta y \, \Delta v_y \geq h/m$$
$$\Delta z \, \Delta v_z \geq h/m \qquad (1),$$

wenn Δ der Unschärfebereich ist und m die Masse des jeweiligen Teilchens. h ist eine Naturkonstante (Planck'sche Konstante), die sehr klein ist. Wir erkennen aus Gl (1), daß mit kleiner werdender Masse die Unschärfe des Teilchens größer wird. D.h.: Das Teilchen beschreibt keine Bahn mehr (das wäre nur für $\Delta x = \Delta v = 0$ der Fall), es bewegt sich akausal!

Bei größeren Massen kann von Bahnbewegungen gesprochen werden, wie wir das im Alltag beobachten, wo wir es mit großen Massensystemen (Ball, Auto) zu tun haben.

Akausal sich bewegende Teilchen können aber nicht durch einen zielgerichteten Schöpfungsakt entstanden sein. D. h.: Es hat nie einen (Schöpfer) Gott gegeben! Der Kosmos ist daher ein Zufallsprodukt! Die akausalen Teilchen können nur durch Wahrscheinlichkeitsaussagen erfaßt werden. Das gilt allerdings auch für größere Teilchen, wenn auch in schwächerem Maße; die Wahrscheinlichkeit ihres Aufenthalts (Aufenthaltswahrscheinlichkeitsdichte $|\underline{\psi}|^2$) tritt lokalisierter auf.

$\underline{\psi}$ selbst (Wellenfunktion) wird als Lösung der Schrödingergleichung erhalten und hängt auch von der Zeit ab, bei vorgegebener Anfangsbedingung des Systems.

Die Gl (1) als unwiderrufliche Erkenntnis (s. o.) hat also zu weitreichenden Konsequenzen geführt, die wir weiterverfolgen wollen.

Nach allem sind die Grundlagen der Religionen nicht mehr zu begründen, wir müssen nun den Glauben als ein Fürwahrhalten erklären; eine Notwendigkeit für Religionen hat nie bestanden. Damit sind auch Seele und Geist (u. a.) neu zu betrachten: Sie stellen Funktionen in ausreichend komplexen Strukturen der Materie dar! Der Tod ist somit das Ende der entsprechenden komplexen Materiestrukturen, ohne Erhalt und Wiederauftreten, die Komplexität schließt dies aus.

Aus allem folgt aber zwingend eine kosmische Ethik für ausreichend komplexe Strukturen im Universum!

Der Urgrund alles Menschlichen ist allen Wesen gemeinsam. Es ist sozusagen ein „materielles Kollektiv“ entstanden – manche Fremdheit zwischen den Systemen kann daher nicht mehr auftreten, wenn diese Erkenntnis erfaßt wird. Damit ist eine gegenseitige Achtung voreinander selbstverständlich geworden, die auch gegenseitige Hilfe einschließt.

Diese Einsicht und Erfassung einer derartigen kosmischen Ethik verlangt auch eine Überprüfung und Korrektur der bisherigen Gesellschaftsstrukturen.

Schließlich eine letzte Konsequenz:

Da die Entstehung und Bildung des Universums kein bewußter Akt (Schöpfung) gewesen sein kann, was aus der Materie spricht (Unschärfe), muß angenommen werden, daß von „außen“ kein Beitrag geleistet wurde. Das heißt wiederum: gesamt betrachtet muß eine Invarianz vorliegen. Der Kosmos ist somit konsequenterweise ein „aufgespaltenes Nichts“ (analog: $+1-1=0$), immer eine „Nullsumme“ insgesamt! Dabei können sich die einzelnen Aufspaltungen ändern entsprechend ihrer eigenen Gesetzmäßigkeit, ohne das „Nichts“ im Gesamten zu verändern.

Die Aufspaltung muß auch die Entstehung der Zeit und des Raumes sein (Existenz und Abstand).

Auch diese Erkenntnis, die, wie gesagt, nur von ausreichend komplexen Materiesystemen erfaßt werden kann, sollte die kosmische Ethik ergänzen und somit in das neue „Weltbild" des Menschen eingebaut werden. Damit sind die wesentlichen Aspekte angedeutet.

Also: Nicht gottlos (Atheist), sondern gottfrei.

Literatur des Autors zu seinem Beitrag:

1) Atome und Moleküle als Bausteine der Materie. Salle-Sauerländer-Verlag. 1982. ISBN 3-7935-5695-6, ISBN 3-7941-2133-3

2) Mensch und Materie. WECHSELWIRKUNGEN. Jahrbuch 1985; Neudruck 1987. Universität Stuttgart

3) Materie ist nicht materiell. Vieweg-Verlag 1997. ISBN 3-528-06666-0. ISSN 0949. Springer-Verlag 2000. ISBN 3-540-67044-0

4) Das Atom und die Ethik. LIT-Verlag 2009. ISBN 978-3-643-10104-4

Quelle: www.Heinzwerner-Preuss.de

About the author

Born on 12 September 1925 in Liegnitz (Niederschlesien).

Studied mathematics beginning in 1945 at the University of Halle/S., took up physics at the University of Hamburg in 1948.

Beginning in 1952, a scientific assistant to Prof. W. Heisenberg at the Max Planck Institute for Physics, first in Göttingen and then later in Munich, where he helped to start the working group on quantum chemistry. In 1969 appointed to the newly founded chair of theoretical chemistry (founding of the institute) at the University of Stuttgart. In these years, he also participated in various academic exchanges in both Sweden and the United States.

Retired (*entpflichtet*) since 1993, October.

The author has published numerous books for specialists. Since 1951, he has more than 200 publications, including work with colleagues. In 1997 his book *Matter is not material* (original title *Materie ist nicht materiell*) was published.

The author remains academically active, particularly in the field of the non-locality of quantum physics.